Douglas H. Campbell

Elements of Structural
and Systematic Botany

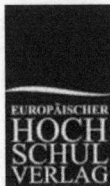

EUROPÄISCHER
HOCH
SCHUL
VERLAG

Douglas H. Campbell

Elements of Structural and Systematic Botany

Reihe: Historical Science, Band 31

1. Auflage 2010 | ISBN: 978-3-86741-219-3

Europäischer Hochschulverlag GmbH & Co KG, Bremen. (www.eh-verlag.de). Alle Rechte vorbehalten.

Die Deutsche Nationalbibliothek verzeichnet diese Publikation in der Deutschen Nationalbibliografie; detaillierte bibliografische Daten sind im Internet über http://www.dnb.d-nb.de abrufbar.

Dieses Buch beruht auf einem alten Original. Der Verlag hat jedoch am ursprünglichen Text einige geringfügige Veränderungen vorgenommen, um die Übersichtlichkeit und Lesbarkeit zu verbessern.

Douglas H. Campbell

Elements of Structural and Systematic Botany

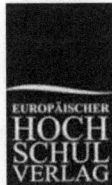

EUROPÄISCHER
HOCH
SCHUL
VERLAG

www.eh-verlag.de

PREFACE.

THE rapid advances made in the science of botany within the last few years necessitate changes in the text books in use as well as in methods of teaching. Having, in his own experience as a teacher, felt the need of a book different from any now in use, the author has prepared the present volume with a hope that it may serve the purpose for which it is intended; viz., an introduction to the study of botany for use in high schools especially, but sufficiently comprehensive to serve also as a beginning book in most colleges.

It does not pretend to be a complete treatise of the whole science, and this, it is hoped, will be sufficient apology for the absence from its pages of many important subjects, especially physiological topics. It was found impracticable to compress within the limits of a book of moderate size anything like a thorough discussion of even the most important topics of *all* the departments of botany. As a thorough understanding of the structure of any organism forms the basis of all further intelligent study of the same, it has seemed to the author proper to emphasize this feature in the present work, which is professedly an *introduction*, only, to the science.

This structural work has been supplemented by so much classification as will serve to make clear the relationships of different groups, and the principles upon which the classification is based, as well as enable the student to recognize the commoner types of the different groups as they are met with. The aim of this book is not, however, merely the identification of plants. We wish here to enter a strong protest against the only too prevalent idea that the chief aim of botany is the ability to run down a plant by means of an "Analytical Key," the subject being exhausted as soon as the name of the plant is discovered. A knowledge of the plant itself is far more important than its name, however desirable it may be to know the latter.

In selecting the plants employed as examples of the different groups, such were chosen, as far as possible, as are everywhere common. Of course this was not always possible, as some important forms, *e.g.* the red and brown seaweeds, are necessarily not always readily procurable by all students, but it will be found that the great majority of the forms used, or closely related ones, are within the reach of nearly all students; and such directions are given for collecting and preserving them as will make it possible even for those in the larger cities to supply themselves with the necessary materials. Such directions, too, for the manipulation and examination of specimens are given as will make the book, it is hoped, a laboratory guide as well as a manual of classification. Indeed, it is primarily intended that the book should so serve as a help in the study of the actual specimens.

Although much can be done in the study, even of the lowest plants, without microscopic aid other than a hand lens, for a thorough understanding of the structure of any plant a good compound microscope is indispensable, and wherever it is possible the student should be provided with such an instrument, to use this book to the best advantage. As, however, many are not able to have the use of a microscope, the gross anatomy of all the forms described has been carefully treated for the especial benefit of such students. Such portions of the text, as well as the general discussions, are printed in ordinary type, while the minute anatomy, and all points requiring microscopic aid, are discussed in separate paragraphs printed in smaller type.

The drawings, with very few exceptions, which are duly credited, were drawn from nature by the author, and nearly all expressly for this work.

A list of the most useful books of reference is appended, all of which have been more or less consulted in the preparation of the following pages.

The classification adopted is, with slight changes, that given in Goebel's "Outlines of Morphology and Classification"; while, perhaps, not in all respects entirely satisfactory, it seems to represent more nearly than any other our present knowledge of the subject. Certain groups, like the Diatoms and *Characeæ*, are puzzles to the botanist, and at present it is impossible to give them more than a provisional place in the system.

If this volume serves to give the student some comprehension of the real aims of botanical science, and its claims to be something more than the "Analysis" of flowers, it will have fulfilled its mission.

DOUGLAS H. CAMPBELL.

BLOOMINGTON, INDIANA,
October, 1889.

TABLE OF CONTENTS.

TABLE OF FIGURES.

CHAPTER I.
INTRODUCTION.

ALL matter is composed of certain constituents (about seventy are at present known), which, so far as the chemist is concerned, are indivisible, and are known as elements.

Of the innumerable combinations of these elements, two general classes may be recognized, organic and inorganic bodies. While it is impossible, owing to the dependence of all organized matter upon inorganic matter, to give an absolute definition, we at once recognize the peculiarities of organic or living bodies as distinguished from inorganic or non-living ones. All living bodies feed, grow, and reproduce, these acts being the result of the action of forces resident within the organism. Inorganic bodies, on the other hand, remain, as a rule, unchanged so long as they are not acted upon by external forces.

All living organisms are dependent for existence upon inorganic matter, and sooner or later return these elements to the sources whence they came. Thus, a plant extracts from the earth and air certain inorganic compounds which are converted by the activity of the plant into a part of its own substance, becoming thus incorporated into a living organism. After the plant dies, however, it undergoes decomposition, and the elements are returned again to the earth and atmosphere from which they were taken.

Investigation has shown that living bodies contain comparatively few elements, but these are combined into extraordinarily complex compounds. The following elements appear to be essential to all living bodies: carbon, hydrogen, oxygen, nitrogen, sulphur, potassium. Besides these there are several others usually present, but not apparently essential to all organisms. These include phosphorus, iron, calcium, sodium, magnesium, chlorine, silicon.

As we examine more closely the structure and functions of organic bodies, an extraordinary uniformity is apparent in all of them. This is disguised in the more specialized forms, but in the simpler ones is very apparent. Owing to this any attempt to separate absolutely the animal and vegetable kingdoms proves futile.

The science that treats of living things, irrespective of the distinction between plant and animal, is called "Biology," but for many purposes it is desirable to recognize the distinctions, making two departments of Biology,—Botany, treating of plants; and Zoölogy, of animals. It is with the first of these only that we shall concern ourselves here.

When one takes up a plant his attention is naturally first drawn to its general appearance and structure, whether it is a complicated one like one of the flowering plants, or some humbler member of the vegetable kingdom,—a moss, seaweed, toadstool,—or even some still simpler plant like a mould, or the apparently structureless green scum that floats on a stagnant pond. In any case the impulse is to investigate the form and structure as far as the means at one's disposal will permit. Such a study of structure constitutes "Morphology," which includes two departments,—gross anatomy, or a general study of the parts; and minute anatomy, or "Histology," in which a microscopic examination is made of the structure of the different parts. A special department of Morphology called "Embryology" is often recognized. This embraces a study of the development of the organism from its earliest stage, and also the development of its different members.

From a study of the structure of organisms we get a clue to their relationships, and upon the basis of such relationships are enabled to classify them or unite them into groups so as to indicate the degree to which they are related. This constitutes the division of Botany usually known as Classification or "Systematic Botany."

Finally, we may study the functions or workings of an organism: how it feeds, breathes, moves, reproduces. This is "Physiology," and like classification must be preceded by a knowledge of the structures concerned.

For the study of the gross anatomy of plants the following articles will be found of great assistance: 1. a sharp knife, and for more delicate tissues, a razor; 2. a pair of small, fine-pointed scissors; 3. a pair of mounted needles (these can be made by forcing ordinary sewing needles into handles of pine or other soft wood); 4. a hand lens; 5. drawing-paper and pencil, and a note book.

For the study of the lower plants, as well as the histology of the higher ones, a compound microscope is indispensable. Instruments with lenses magnifying from about 20 to 500 diameters can be had at a cost varying from about $20 to $30, and are sufficient for any ordinary investigations.

Objects to be studied with the compound microscope are usually examined by transmitted light, and must be transparent enough to allow the light to pass through. The objects are placed upon small glass slips (slides), manufactured for the purpose, and covered with extremely thin plates of glass, also specially made. If the body to be examined is a large one, thin slices or sections must be made. This for most purposes may be done with an ordinary razor. Most plant tissues are best examined ordinarily in water, though of course specimens so mounted cannot be preserved for any length of time. [1]

In addition to the implements used in studying the gross anatomy, the following will be found useful in histological work: 1. a small camel's-hair brush for picking up small sections and putting water in the slides; 2. small forceps for handling delicate objects; 3. blotting paper for removing superfluous water from the slides and drawing fluids under the cover glass; 4. pieces of elder or sunflower pith, for holding small objects while making sections.

In addition to these implements, a few reagents may be recommended for the simpler histological work. The most important of these are alcohol, glycerine, potash (a strong solution of potassium hydrate in water), iodine (either a little of the commercial tincture of iodine in water, or, better, a solution of iodine in iodide of potassium), acetic acid, and some staining fluid. (An aqueous or alcoholic solution of gentian violet or methyl violet is one of the best.)

A careful record should be kept by the student of all work done, both by means of written notes and drawings. For most purposes pencil drawings are most convenient, and these should be made with a moderately soft pencil on unruled paper. If it is desired to make the drawings with ink, a careful outline should first be made with a hard pencil and this inked over with India-ink or black drawing ink. Ink drawings are best made upon light bristol board with a hard, smooth-finished surface.

When obtainable, the student will do best to work with freshly gathered specimens; but as these are not always to be had when wanted, a few words about gathering and preserving material may be of service.

Most of the lower green plants (*algæ*) may be kept for a long time in glass jars or other vessels, provided care is taken to remove all dead specimens at first and to renew the water from time to time. They

3

usually thrive best in a north window where they get little or no direct sunshine, and it is well to avoid keeping them too warm.

Numbers of the most valuable fungi — *i.e.* the lower plants that are not green — grow spontaneously on many organic substances that are kept warm and moist. Fresh bread kept moist and covered with a glass will in a short time produce a varied crop of moulds, and fresh horse manure kept in the same way serves to support a still greater number of fungi.

Mosses, ferns, etc., can be raised with a little care, and of course very many flowering plants are readily grown in pots.

Most of the smaller parasitic fungi (rusts, mildews, etc.) may be kept dry for any length of time, and on moistening with a weak solution of caustic potash will serve nearly as well as freshly gathered specimens for most purposes.

When it is desired to preserve as perfectly as possible the more delicate plant structures for future study, strong alcohol is the best and most convenient preserving agent. Except for loss of color it preserves nearly all plant tissues perfectly.

CHAPTER II.
THE CELL.

IF we make a thin slice across the stem of a rapidly growing plant,—*e.g.* geranium, begonia, celery,—mount it in water, and examine it microscopically, it will be found to be made up of numerous cavities or chambers separated by delicate partitions. Often these cavities are of sufficient size to be visible to the naked eye, and examined with a hand lens the section appears like a piece of fine lace, each mesh being one of the chambers visible when more strongly magnified. These chambers are known as "cells," and of them the whole plant is built up.

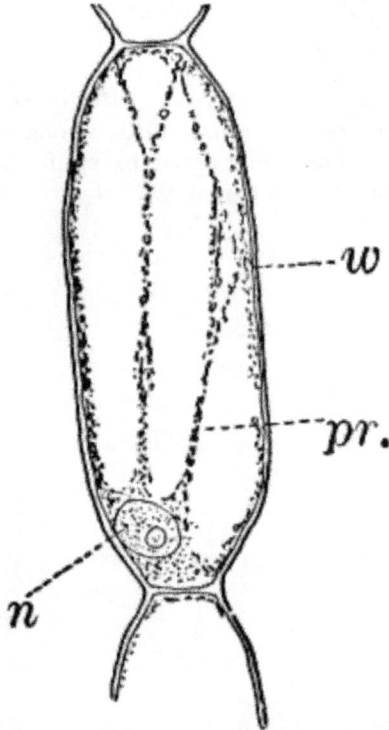

FIG. 1.—*A single cell from a hair on the stamen of the common spiderwort (Tradescantia), × 150. pr. protoplasm; w, cell wall; n, nucleus.*

In order to study the structure of the cell more exactly we will select such as may be examined without cutting them. A good example is furnished by the common spiderwort (Fig. 1). Attached to the base of the stamens (Fig. 85, *B*) are delicate hairs composed of chains of cells, which may be examined alive by carefully removing a stamen and placing it in a drop of water under a cover glass. Each cell (Fig. 1) is an oblong sac, with a delicate colorless wall which chemical tests show to be composed of cellulose, a substance closely resembling starch. Within this sac, and forming a lining to it, is a thin layer of colorless matter containing many fine granules. Bands and threads of the same substance traverse the cavity of the cell, which is filled with a deep purple homogeneous fluid. This fluid, which in most cells is colorless, is called the cell sap, and is composed mainly of water. Imbedded in the granular lining of the sac is a roundish body (*n*), which itself has a definite membrane, and usually shows one or more roundish bodies within, besides an indistinctly granular appearance. This body is called the nucleus of the cell, and the small one within it, the nucleolus.

The membrane surrounding the cell is known as the cell wall, and in young plant cells is always composed of cellulose.

The granular substance lining the cell wall (Fig. 1, *pr.*) is called "protoplasm," and with the nucleus constitutes the living part of the cell. If sufficiently magnified, the granules within the protoplasm will be seen to be in active streaming motion. This movement, which is very evident here, is not often so conspicuous, but still may often be detected without difficulty.

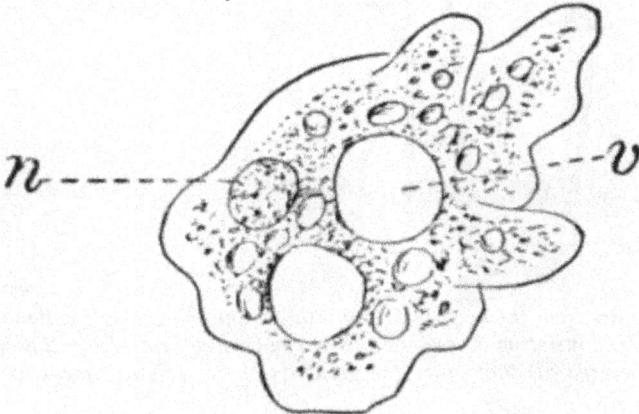

FIG. 2. — *An Amœba. A cell without a cell wall. n, nucleus; v, vacuoles,* × *300.*

6

The cell may be regarded as the unit of organic structure, and of cells are built up all of the complicated structures of which the bodies of the highest plants and animals are composed. We shall find that the cells may become very much modified for various purposes, but at first they are almost identical in structure, and essentially the same as the one we have just considered.

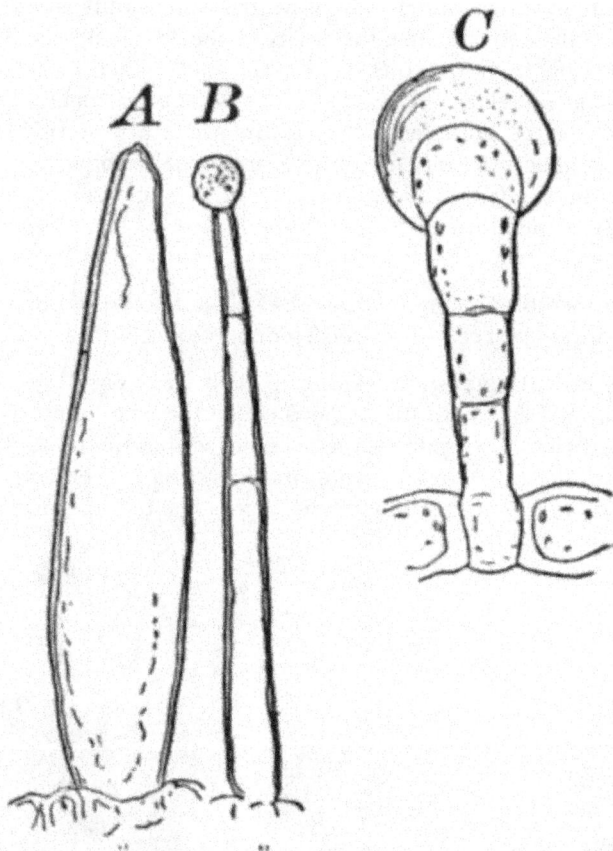

FIG. 3. — *Hairs from the leaf stalk of a wild geranium. A, single-celled hair. B and C, hairs consisting of a row of cells. The terminal rounded cell secretes a peculiar scented oil that gives the plant its characteristic odor. B, × 50; C, × 150.*

Very many of the lower forms of life consist of but a single cell which may occasionally be destitute of a cell wall. Such a form is shown in Figure 2. Here we have a mass of protoplasm with a nucleus (*n*) and cavities (vacuoles, *v*) filled with cell sap, but no cell wall. The protoplasm is in constant movement, and by extensions of a portion of the mass and contraction of other parts, the whole creeps slowly along. Other naked cells (Fig. 12, *B*; Fig. 16, *C*) are provided with delicate thread-like processes of protoplasm called "cilia" (sing. *cilium*), which are in active vibration, and propel the cell through the water.

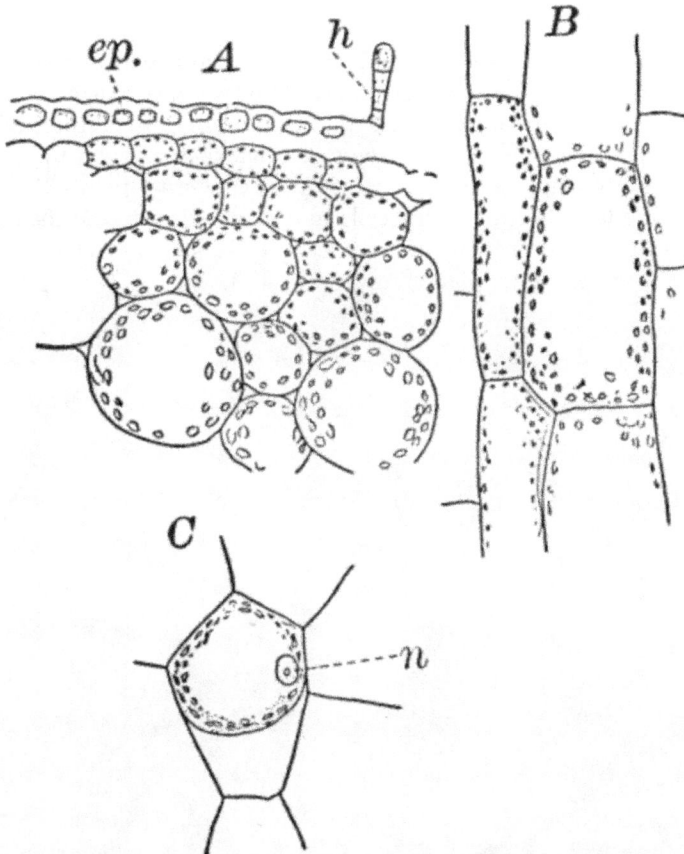

Fig. 4.—A, cross section. B, longitudinal section of the leaf stalk of wild geranium, showing its cellular structure. Ep. epidermis. h, a hair, × 50. C, a cell from the prothallium (young plant) of a fern, × 150. The contents of the cell contracted by the action of a solution of sugar.

On placing a cell into a fluid denser than the cell sap (*e.g.* a ten-per-cent solution of sugar in water), a portion of the water will be extracted from the cell, and we shall then see the protoplasm receding from the wall (Fig. 4, *C*), showing that it is normally in a state of tension due to pressure from within of the cell sap. The cell wall shows the same thing though in a less degree, owing to its being much more rigid than the protoplasmic lining. It is owing to the partial collapsing of the cells, consequent on loss of water, that plants wither when the supply of water is cut off.

As cells grow, new ones are formed in various ways. If the new cells remain together, cell aggregates, called tissues, are produced, and of these tissues are built up the various organs of the higher plants. The simplest tissues are rows of cells, such as form the hairs covering the surface of the organs of many flowering plants (Fig. 3), and are due to a division of the cells in a single direction. If the divisions take place in three planes, masses of cells, such as make up the stems, etc., of the higher plants, result (Fig. 4, *A*, *B*).

CHAPTER III.
CLASSIFICATION OF PLANTS. – PROTOPHYTES.

FOR the sake of convenience it is desirable to collect into groups such plants as are evidently related; but as our knowledge of many forms is still very imperfect, any classification we may adopt must be to a great extent only provisional, and subject to change at any time, as new forms are discovered or others become better understood.

The following general divisions are usually accepted: I. Sub-kingdom (or Branch); II. Class; III. Order; IV. Family; V. Genus; VI. Species.

To illustrate: The white pine belongs to the highest great division (sub-kingdom) of the plant kingdom. The plants of this division all produce seeds, and hence are called "spermaphytes" ("seed plants"). They may be divided into two groups (classes), distinguished by certain peculiarities in the flowers and seeds. These are named respectively "gymnosperms" and "angiosperms," and to the first our plant belongs. The gymnosperms may be further divided into several subordinate groups (orders), one of which, the conifers, or cone-bearing evergreens, includes our plant. This order includes several families, among them the fir family (*Abietineæ*), including the pines and firs. Of the sub-divisions (*genera*, sing. *genus*) of the fir family, one of the most familiar is the genus *Pinus*, which embraces all the true pines. Comparing different kinds of pines, we find that they differ in the form of the cones, arrangement of the leaves, and other minor particulars. The form we have selected differs from all other native forms in its cones, and also in having the leaves in fives, instead of twos or threes, as in most other kinds. Therefore to distinguish the white pine from all other pines, it is given a "specific" name, *strobus*.

The following table will show more plainly what is meant:

Sub-kingdom,

Spermaphyta.

Includes all spermaphytes, or seed plants.

Class,

Gymnospermæ.

All naked-seeded plants.

Order,

Coniferæ.

All cone-bearing evergreens.

Family,

Abietineæ.

Firs, Pines, etc.

Genus,

Pinus.

Pines.

Species,

Strobus.

White Pine.

SUB-KINGDOM I.
Protophytes.

The name Protophytes (*Protophyta*) has been applied to a large number of simple plants, which differ a good deal among themselves. Some of them differ strikingly from the higher plants, and resemble so remarkably certain low forms of animal life as to be quite indistinguishable from them, at least in certain stages. Indeed, there are certain forms that are quite as much animal as vegetable in their attributes, and must be regarded as connecting the two kingdoms. Such forms are the slime moulds (Fig. 5), *Euglena* (Fig. 9), *Volvox* (Fig. 10), and others.

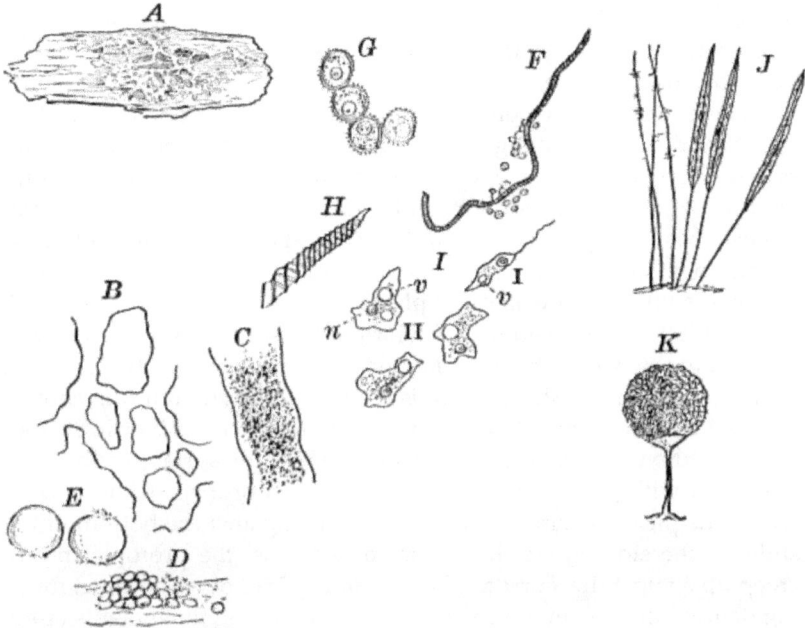

Fig. 5.—A, a portion of a slime mould growing on a bit of rotten wood, × 3. B, outline of a part of the same, × 25. C, a small portion showing the densely granular character of the protoplasm, × 150. D, a group of spore cases of a slime mould (Trichia), of about the natural size. E, two spore cases, × 5. The one at the right has begun to open. F, a thread (capillitium) and spores of Trichia, × 50. G, spores. H, end of the thread, × 300. I, zoöspores of Trichia, × 300. i, ciliated form; ii, amœboid forms. n, nucleus. v, contractile vacuole. J, K, sporangia of two common slime moulds. J, Stemonitis, × 2. K, Arcyria, × 4.

Other protophytes, while evidently enough of vegetable nature, are nevertheless very different in some respects from the higher plants.

The protophytes may be divided into three classes: I. The slime moulds (*Myxomycetes*); II. The Schizophytes; III. The green monads (*Volvocineæ*).

Class I. — The Slime Moulds.

These curious organisms are among the most puzzling forms with which the botanist has to do, as they are so much like some of the lowest forms of animal life as to be scarcely distinguishable from them, and indeed they are sometimes regarded as animals rather than plants. At certain stages they consist of naked masses of protoplasm of very considerable size, not infrequently several centimetres in diameter. These are met with on decaying logs in damp woods, on rotting leaves, and other decaying vegetable matter. The commonest ones are bright yellow or whitish, and form soft, slimy coverings over the substratum (Fig. 5, *A*), penetrating into its crevices and showing sensitiveness toward light. The plasmodium, as the mass of protoplasm is called, may be made to creep upon a slide in the following way: A tumbler is filled with water and placed in a saucer filled with sand. A strip of blotting paper about the width of the slide is now placed with one end in the water, the other hanging over the edge of the glass and against one side of a slide, which is thus held upright, but must not be allowed to touch the side of the tumbler. The strip of blotting paper sucks up the water, which flows slowly down the surface of the slide in contact with the blotting paper. If now a bit of the substance upon which the plasmodium is growing is placed against the bottom of the slide on the side where the stream of water is, the protoplasm will creep up against the current of water and spread over the slide, forming delicate threads in which most active streaming movements of the central granular protoplasm may be seen under the microscope, and the ends of the branches may be seen to push forward much as we saw in the amœba. In order that the experiment may be successful, the whole apparatus should be carefully protected from the light, and allowed to stand for several hours. This power of movement, as well as the power to take in solid food, are eminently animal characteristics, though the former is common to many plants as well.

After a longer or shorter time the mass of protoplasm contracts and gathers into little heaps, each of which develops into a structure that has no resemblance to any animal, but would be at once placed with plants. In one common form (*Trichia*) these are round or pear-shaped bodies of a yellow color, and about as big as a pin head (Fig. 5, *D*), occurring in groups on rotten logs in damp woods. Others are stalked (*Arcyria, Stemonitis*) (Fig. 5, *J, K*), and of various colors, — red, brown, etc. The outer part of the structure is a more or less firm wall, which breaks when ripe, discharging a powdery mass, mixed in most forms with very fine fibres.

When strongly magnified the fine dust is found to be made up of innumerable small cells with thick walls, marked with ridges or processes which differ much in different species. The fibres also differ much in different genera. Sometimes they are simple, hair-like threads; in others they are hollow tubes with spiral thickenings, often very regularly placed, running around their walls.

The spores may sometimes be made to germinate by placing them in a drop of water, and allowing them to remain in a warm place for about twenty-four hours. If the experiment has been successful, at the end of this time the spore membrane will have burst, and the contents escaped in the form of a naked mass of protoplasm (Zoöspore) with a nucleus, and often showing a vacuole (Fig. 5, *v*), that alternately becomes much distended, and then disappears entirely. On first escaping it is usually provided with a long, whip-like filament of protoplasm, which is in active movement, and by means of which the cell swims actively through the water (Fig. 5, *I* I). Sometimes such a cell will be seen to divide into two, the process taking but a short time, so that the numbers of these cells under favorable conditions may become very large. After a time the lash is withdrawn, and the cell assumes much the form of a small amœba (*I* II).

The succeeding stages are difficult to follow. After repeatedly dividing, a large number of these amœba-like cells run together, coalescing when they come in contact, and forming a mass of protoplasm that grows, and finally assumes the form from which it started.

Of the common forms of slime moulds the species of *Trichia* (Figs. *D*, *I*) and *Physarum* are, perhaps, the best for studying the germination, as the spores are larger than in most other forms, and germinate more readily. The experiment is apt to be most successful if the spores are sown in a drop of water in which has been infused some vegetable matter, such as a bit of rotten wood, boiling thoroughly to kill all germs. A drop of this fluid should be placed on a perfectly clean cover glass, which it is well to pass once or twice through a flame, and the spores transferred to this drop with a needle previously heated. By these precautions foreign germs will be avoided, which otherwise may interfere seriously with the growth of the young slime moulds. After sowing the spores in the drop of culture fluid, the whole should be inverted over a so-called "moist chamber." This is simply a square of thick blotting paper, in which an opening is cut small enough to be entirely covered by the cover glass, but large enough so that the drop in the centre of the cover glass will not touch the sides of the chamber, but will hang suspended clear in it. The blotting paper should be soaked thoroughly in pure water (distilled water is preferable), and then placed on a slide, covering carefully with the cover glass with the suspended drop of fluid containing the spores. The whole should be kept under cover so as to prevent loss of water by evaporation. By this method the spores may be examined conveniently without disturbing them, and the whole may be kept as long as desired, so long as the blotting paper is kept wet, so as to prevent the suspended drop from drying up.

Class II. — Schizophytes.

The Schizophytes are very small plants, though not infrequently occurring in masses of considerable size. They are among the commonest of all plants, and are found everywhere. They multiply almost entirely by simple transverse division, or splitting of the cells, whence their name. There are two pretty well-marked orders, — the blue-green slimes (*Cyanophyceæ*) and the bacteria (*Schizomycetes*). They are distinguished, primarily, by the first (with a very few exceptions) containing chlorophyll (leaf-green), which is entirely absent from nearly all of the latter.

The blue-green slimes: These are, with few exceptions, green plants of simple structure, but possessing, in addition to the ordinary green pigment (chlorophyll, or leaf-green), another coloring matter, soluble in water, and usually blue in color, though sometimes yellowish or red.

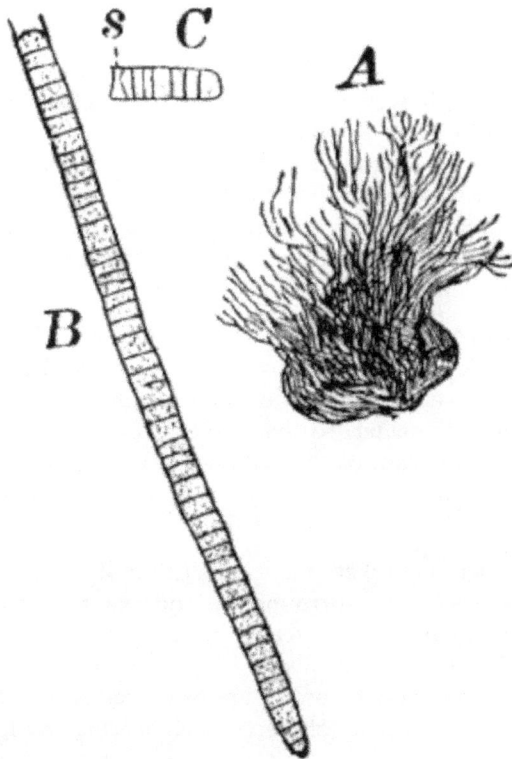

FIG. 6.—Blue-green slime (Oscillaria). A, mass of filaments of the natural size. B, single filament, × 300. C, a piece of a filament that has become separated. s, sheath, × 300.

As a representative of the group, we will select one of the commonest forms (*Oscillaria*), known sometimes as green slime, from forming a dark blue-green or blackish slimy coat over the mud at the bottom of stagnant or sluggish water, in watering troughs, on damp rocks, or even on moist earth. A search in the places mentioned can hardly fail to secure plenty of specimens for study. If a bit of the slimy mass is transferred to a china dish, or placed with considerable water on a piece of stiff paper, after a short time the edge of the mass will show numerous extremely fine filaments of a dark blue-green color, radiating in all directions from the mass (Fig. 6, *a*). The filaments are the individual plants, and possess considerable power of motion, as is shown by letting the mass remain undisturbed for a day or two, at the end of which time they will have formed a thin film over the surface

16

of the vessel in which they are kept; and the radiating arrangement of the filaments can then be plainly seen.

If the mass is allowed to dry on the paper, it often leaves a bright blue stain, due to the blue pigment in the cells of the filament. This blue color can also be extracted by pulverizing a quantity of the dried plants, and pouring water over them, the water soon becoming tinged with a decided blue. If now the water containing the blue pigment is filtered, and the residue treated with alcohol, the latter will extract the chlorophyll, becoming colored of a yellow-green.

The microscope shows that the filaments of which the mass is composed (Fig. 6, B) are single rows of short cylindrical cells of uniform diameter, except at the end of the filament, where they usually become somewhat smaller, so that the tip is more or less distinctly pointed. The protoplasm of the cells has a few small granules scattered through it, and is colored uniformly of a pale blue-green. No nucleus can be seen.

If the filament is broken, there may generally be detected a delicate, colorless sheath that surrounds it, and extends beyond the end cells (Fig. 6, c). The filament increases in length by the individual cells undergoing division, this always taking place at right angles to the axis of the filament. New filaments are produced simply by the older ones breaking into a number of pieces, each of which rapidly grows to full size.

The name "oscillaria" arises from the peculiar oscillating or swinging movements that the plant exhibits. The most marked movement is a swaying from side to side, combined with a rotary motion of the free ends of the filaments, which are often twisted together like the strands of a rope. If the filaments are entirely free, they may often be observed to move forward with a slow, creeping movement. Just how these movements are caused is still a matter of controversy.

The lowest of the *Cyanophyceæ* are strictly single-celled, separating as soon as formed, but cohering usually in masses or colonies by means of a thick mucilaginous substance that surrounds them (Fig. 7, D).

The higher ones are filaments, in which there may be considerable differentiation. These often occur in masses of considerable size, forming jelly-like lumps, which may be soft or quite firm (Fig. 7, A, B). They are sometimes found on damp ground, but more commonly attached to plants, stones, etc., in water. The masses vary in color from light brown to deep blackish green, and in size from that of a pin head to several centimetres in diameter.

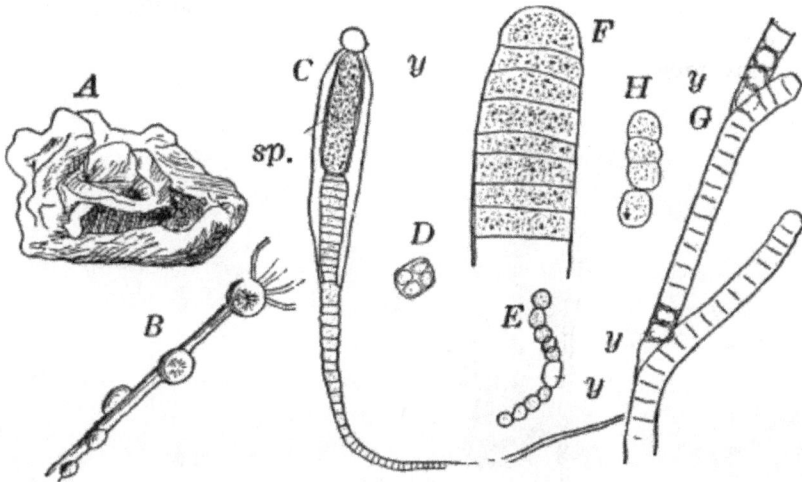

Fig. 7.—Forms of Cyanophyceæ. A, Nostoc. B, Glæotrichia, × 1. C, individual of Glæotrichia. D, Chroöcoccus. E, Nostoc. F, Oscillaria. G, H, Tolypothrix. All × 300. y, heterocyst. sp. spore.

In the higher forms special cells called heterocysts are found. They are colorless, or light yellowish, regularly disposed; but their function is not known. Besides these, certain cells become thick-walled, and form resting cells (spores) for the propagation of the plant (Fig. 7, C. *sp.*). In species where the sheath of the filament is well marked (Fig. 7, *H*), groups of cells slip out of the sheath, and develop a new one, thus giving rise to a new plant.

The bacteria (*Schizomycetes*), although among the commonest of organisms, owing to their excessive minuteness, are difficult to study, especially for the beginner. They resemble, in their general structure and methods of reproduction, the blue-green slimes, but are, with very few exceptions, destitute of chlorophyll, although often possessing bright pigments,—blue, violet, red, etc. It is one of these that sometimes forms blood-red spots in flour paste or bits of bread that have been kept very moist and warm. They are universally present where decomposition is going on, and are themselves the principal agents of decay, which is the result of their feeding upon the substance, as, like all plants without chlorophyll, they require organic matter for food. Most of the species are very tenacious of life, and may be completely dried up for a long time without dying, and on being placed in water will quickly revive. Being so extremely small, they are readily carried about in the air in their dried-up condition, and thus fall upon exposed bodies, setting up decomposition if the conditions are favorable.

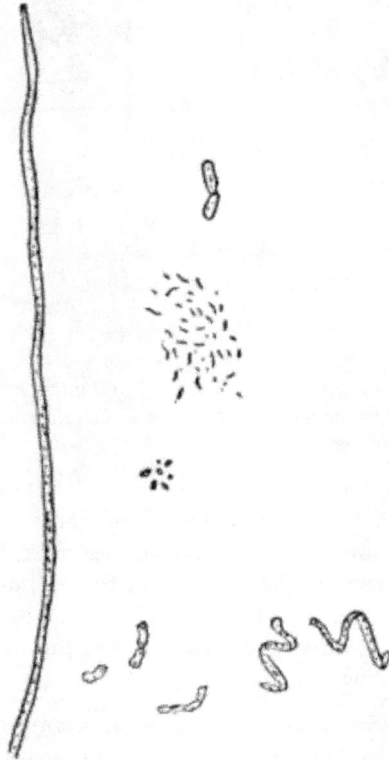

FIG. 8.—Bacteria.

A simple experiment to show this may be performed by taking two test tubes and partly filling them with an infusion of almost any organic substance (dried leaves or hay, or a bit of meat will answer). The fluid should now be boiled so as to kill any germs that may be in it; and while hot, one of the vessels should be securely stopped up with a plug of cotton wool, and the other left open. The cotton prevents access of all solid particles, but allows the air to enter. If proper care has been taken, the infusion in the closed vessel will remain unchanged indefinitely; but the other will soon become turbid, and a disagreeable odor will be given off. Microscopic examination shows the first to be free from germs of any kind, while the second is swarming with various forms of bacteria.

These little organisms have of late years attracted the attention of very many scientists, from the fact that to them is due many, if not all, contagious diseases. The germs of many such diseases have been isolated, and experiments prove beyond doubt that these are alone the causes of the diseases in question.

If a drop of water containing bacteria is examined, we find them to be excessively small, many of them barely visible with the strongest lenses. The larger ones (Fig. 8) recall quite strongly the smaller species of oscillaria, and exhibit similar movements. Others are so small as to appear as mere lines and dots, even with the strongest lenses. Among the common forms are small, nearly globular cells; oblong, rod-shaped or thread-shaped filaments, either straight or curved, or even spirally twisted. Frequently they show a quick movement which is probably in all cases due to cilia, which are, however, too small to be seen in most cases.

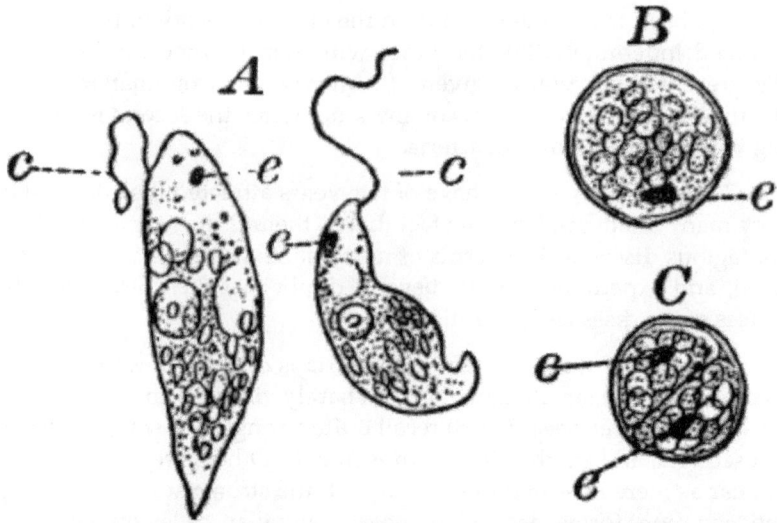

FIG. 9.—*Euglena. A, individual in the active condition. E, the red "eye-spot." c, flagellum. n, nucleus. B, resting stage. C, individual dividing, × 300.*

Reproduction is for the most part by simple transverse division, as in oscillaria; but occasionally spores are produced also.

Class III. — Green Monads (*VOLVOCINEÆ*).

This group of the protophytes is unquestionably closely related to certain low animals (*Monads* or *Flagellata*), with which they are sometimes united. They are characterized by being actively motile, and are either strictly unicellular, or the cells are united by a gelatinous envelope into a colony of definite form.

Of the first group, *Euglena* (Fig. 9), may be selected as a type.

This organism is found frequently among other algæ, and occasionally forms a green film on stagnant water. It is sometimes regarded as a plant, sometimes as an animal, and is an elongated, somewhat worm-like cell without a definite cell wall, so that it can change its form to some extent. The protoplasm contains oval masses, which are bright green in color; but the forward pointed end of the cell is colorless, and has a little depression. At this end there is a long vibratile protoplasmic filament (*c*), by means of which the cell moves. There is also to be seen near this end a red speck (*e*) which is probably sensitive to light. A nucleus can usually be seen if the cell is first killed with an iodine solution, which often will render the flagellum (*c*) more evident, this being invisible while the cell is in motion. The cells multiply by division. Previous to this the flagellum is withdrawn, and a firm cell wall is formed about the cell (Fig. 9, *B*). The contents then divide into two or more parts, which afterwards escape as new individuals.

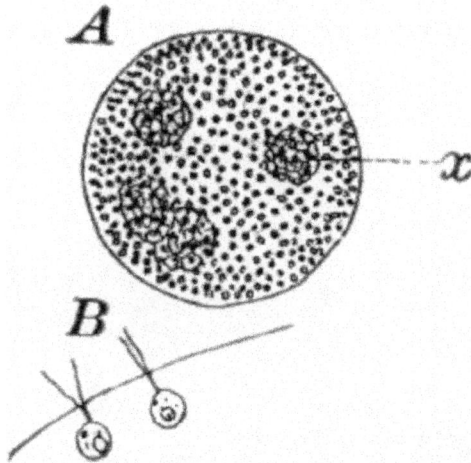

FIG. 10. — *Volvox. A, mature colony, containing several smaller ones (x), × 50. B, Two cells showing the cilia, × 300.*

Of the forms that are united in colonies [2] one of the best known is *Volvox* (Fig. 10). This plant is sometimes found in quiet water, where it floats on or near the surface as a dark green ball, just large enough to be seen with the naked eye. They may be kept for some time in aquaria, and will sometimes multiply rapidly, but are very susceptible to extremes of temperature, especially of heat.

The colony (Fig. 10, *A*) is a hollow sphere, the numerous green cells of which it is composed forming a single layer on the outside. By killing with iodine, and using a strong lens, each cell is seen to be somewhat pear-shaped (Fig. *B*), with the pointed end out. Attached to this end are two vibratile filaments (cilia or *flagella*), and the united movements of these cause the rolling motion of the whole colony. Usually a number of young colonies (Fig. *x*) are found within the mother colony. These arise by the repeated bipartition of a single cell, and escape finally, forming independent colonies.

Another (sexual) form of reproduction occurs, similar to that found in many higher plants; but as it only occurs at certain seasons, it is not likely to be met with by the student.

Other forms related to *Volvox*, and sometimes met with, are *Gonium*, in which there are sixteen cells, forming a flat square; *Pandorina* and *Eudorina*, with sixteen cells, forming an oval or globular colony like *Volvox*, but much smaller. In all of these the structure of the cells is essentially as in *Volvox*.

CHAPTER IV.
SUB-KINGDOM II.
ALGÆ. [3]

IN the second sub-kingdom of plants is embraced an enormous assemblage of plants, differing widely in size and complexity, and yet showing a sufficiently complete gradation from the lowest to the highest as to make it impracticable to make more than one sub-kingdom to include them. They are nearly all aquatic forms, although many of them will survive long periods of drying, such forms occurring on moist earth, rocks, or the trunks of trees, but only growing when there is a plentiful supply of water.

All of them possess chlorophyll, which, however, in many forms, is hidden by the presence of a brown or red pigment. They are ordinarily divided into three classes—I. The Green Algæ (*Chlorophyceæ*); II. Brown Algæ (*Phæophyceæ*); III. Red Algæ (*Rhodophyceæ*).

Class I.—Green Algæ.

The green algæ are to be found almost everywhere where there is moisture, but are especially abundant in sluggish or stagnant fresh water, being much less common in salt water. They are for the most part plants of simple structure, many being unicellular, and very few of them plants of large size.

We may recognize five well-marked orders of the green algæ— I. Green slimes (*Protococcaceæ*); II. *Confervaceæ*; III. Pond scums (*Conjugatæ*); IV. *Siphoneæ*; V. Stone-worts (*Characeæ*).

Order I.—Protococcaceæ.

The members of this order are minute unicellular plants, growing either in water or on the damp surfaces of stones, tree trunks, etc. The plants sometimes grow isolated, but usually the cells are united more or less regularly into colonies.

A common representative of the order is the common green slime, *Protococcus* (Fig. 11, *A*, *C*), which forms a dark green slimy coating over stones, tree trunks, flower pots, etc. Owing to their minute size the structure can only be made out with the microscope.

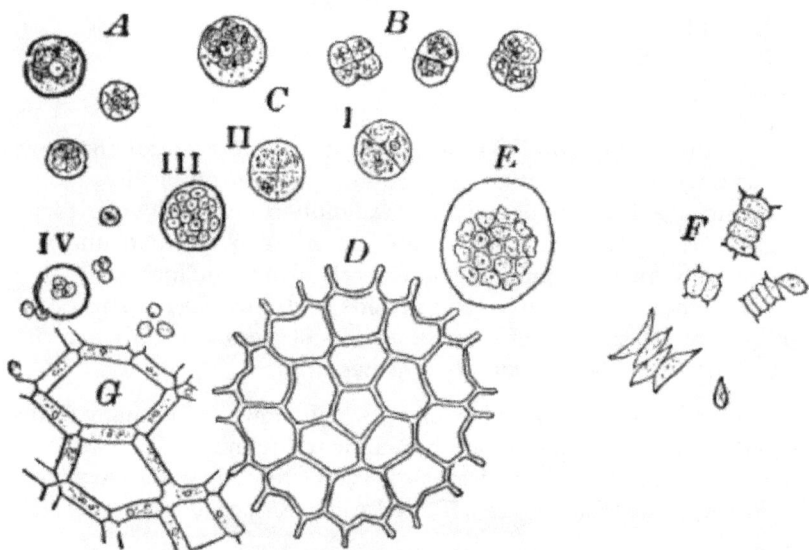

Fig. 11.—Protococcaceæ. A, C, Protococcus. A, single cells. B, cells dividing by fission. C, successive steps in the process of internal cell division. In C iv, the young cells have mostly become free. D, a full-grown colony of Pedias-trum. E, a young colony still surrounded by the membrane of the mother cell. F, Scenedesmus. All, × 300. G, small portion of a young colony of the water net (Hydrodictyon), × 150.

Scraping off a little of the material mentioned into a drop of water upon a slide, and carefully separating it with needles, a cover glass may be placed over the preparation, and it is ready for examination. When magnified, the green film is found to be composed of minute globular cells of varying size, which may in places be found to be united into groups. With a higher power, each cell (Fig. 11, A) is seen to have a distinct cell wall, within which is colorless protoplasm. Careful examination shows that the chlorophyll is confined to several roundish bodies that are not usually in immediate contact with the wall of the cell. These green masses are called chlorophyll bodies (chloroplasts). Toward the centre of the cell, especially if it has first been treated with iodine, the nucleus may be found. The size of the cells, as well as the number of chloroplasts, varies a good deal.

With a little hunting, specimens in various stages of division may be found. The division takes place in two ways. In the first (Fig. 11, B), known as fission, a wall is formed across the cell, dividing it into two cells, which may separate immediately or may remain united until they have undergone further division. In this case the original cell wall remains as part of the wall of the daughter cells. Fission is the commonest form of cell multiplication throughout the vegetable kingdom.

The second form of cell division or internal cell division is shown at C. Here the protoplasm and nucleus repeatedly divide until a number of small cells are formed within the old one. These develop cell walls, and escape by the breaking of the old cell wall, which is left behind, and takes no part in the process. The cells thus formed are sometimes provided with two cilia, and are capable of active movement.

Internal cell division, as we shall see, is found in most plants, but only at special times.

Closely resembling *Protococcus*, and answering quite as well for study, are numerous aquatic forms, such as *Chlorococcum* (Fig. 12). These are for the most part destitute of a firm cell wall, but are imbedded in masses of gelatinous substance like many *Cyanophyceæ*. The chloroplasts are smaller and less distinct than in *Protococcus*. The cells are here oval rather than round, and often show a clear space at one end.

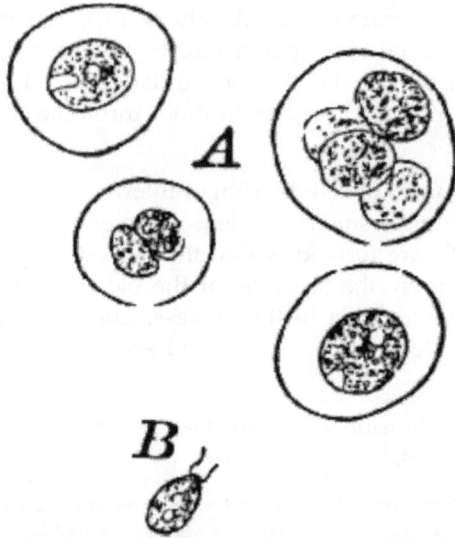

FIG. 12.—*Chlorococcum, a plant related to Protococcus, but the naked cells are surrounded by a colorless gelatinous envelope. A, motionless cells. B, a cell that has escaped from its envelope and is ciliated, × 300.*

Owing to the absence of a definite membrane, a distinction between fission and internal cell division can scarcely be made here. Often the cells escape from the gelatinous envelope, and swim actively by means of two cilia at the colorless end (Fig. 12, *B*). In this stage they closely resemble the individuals of a *Volvox* colony, or other green *Flagellata*, to which there is little doubt that they are related.

There are a number of curious forms common in fresh water that are probably related to *Protococcus*, but differ in having the cells united in colonies of definite form. Among the most striking are the different species of *Pediastrum* (Fig. 11, *D, E*), often met with in company with other algæ, and growing readily in aquaria when once established. They are of very elegant shapes, and the number of cells some multiple of four, usually sixteen.

The cells form a flat disc, the outer ones being generally provided with a pair of spines.

New individuals arise by internal division of the cells, the contents of each forming as many parts as there are cells in the whole colony. The young cells now escape through a cleft in the wall of the mother cell, but are still surrounded by a delicate membrane (Fig. 11, *E*). Within this membrane the young cells arrange themselves in the form of the original colony, and grow together, forming a new colony.

A much larger but rarer form is the water net (Fig. 11, *G*), in which the colony has the form of a hollow net, the spaces being surrounded by long cylindrical cells placed end to end. Other common forms belong to the genus *Scenedesmus* (Fig. 11, *F*), of which there are many species.

Order II. — *CONFERVACEÆ*.

Under this head are included a number of forms of which the simplest ones approach closely, especially in their younger stages, the *Protococcaceæ*. Indeed, some of the so-called *Protococcaceæ* are known to be only the early stages of these plants.

A common member of this order is *Cladophora*, a coarse-branching alga, growing commonly in running water, where it forms tufts, sometimes a metre or more in length. By floating out a little of it in a saucer, it is easy to see that it is made up of branching filaments.

The microscope shows (Fig. 13, *A*) that these filaments are rows of cylindrical cells with thick walls showing evident stratification. At intervals branches are given off, which may in turn branch, giving rise to a complicated branching system. These branches begin as little protuberances of the cell wall at the top of the cell. They increase rapidly in length, and becoming slightly contracted at the base, a wall is formed across at this point, shutting it off from the mother cell.

The protoplasm lines the wall of the cell, and extends in the form of thin plates across the cavity of the cell, dividing it up into a number of irregular chambers. Imbedded in the protoplasm are numerous flattened chloroplasts, which are so close together as to make the protoplasm appear almost uniformly green. Within the chloroplasts are globular, glistening bodies, called "pyrenoids." The cell has several nuclei, but they are scarcely evident in the living cell. By placing the cells for a few hours in a one per cent watery solution of chromic acid, then washing thoroughly and staining with borax carmine, the nuclei will be made very evident (Fig. 13, *B*). Such preparations may be kept permanently in dilute glycerine.

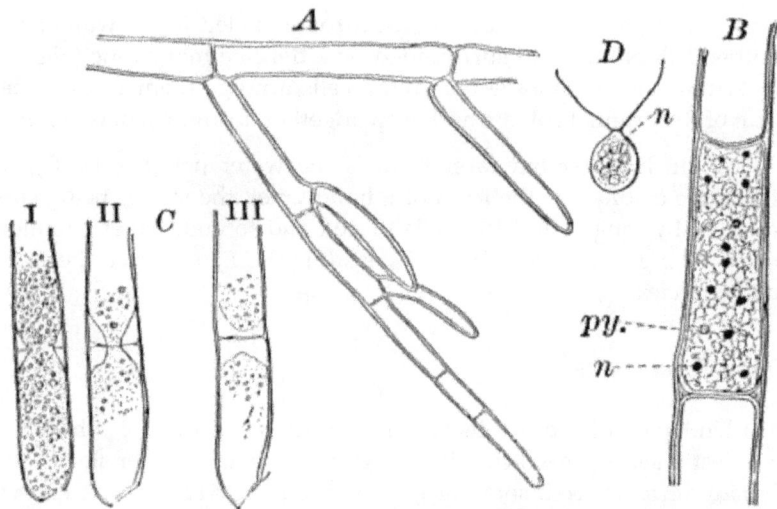

FIG. 13. — *Cladophora. A, a fragment of a plant,* × 50. *B, a single cell treated with chromic acid, and stained with alum cochineal. n, nucleus. py. pyrenoid,* × 150. *C, three stages in the division of a cell.* I, 1.45 p.m.; II, 2.55 p.m.; III, 4.15 p.m., × 150. *D, a zoöspore* × 350.

If a mass of actively growing filaments is examined, some of the cells will probably be found in process of fission. The process is very simple, and may be easily followed (Fig. 13, C). A ridge of cellulose is formed around the cell wall, projecting inward, and pushing in the protoplasm as it grows. The process is continued until the ring closes in the middle, cutting the protoplasmic body completely in two, and forms a firm membrane across the middle of the cell. The protoplasm at this stage (C III.) is somewhat contracted, but soon becomes closely applied to the new wall. The whole process lasts, at ordinary temperatures (20°-25° C.), from three to four hours.

At certain times, but unfortunately not often to be met with, the contents of some of the cells form, by internal division, a large number of small, naked cells (zoöspores) (Fig. 13, D), which escape and swim about actively for a time, and afterwards become invested with a cell wall, and grow into a new filament. These cells are called zoöspores, from their animal-like movements. They are provided with two cilia, closely resembling the motile cells of the *Protococcaceæ* and *Volvocineæ*.

29

There are very many examples of these simple *Confervaceæ*, some like *Conferva* being simple rows of cells, others like *Stigeoclonium* (Fig. 14, *A*), *Chætophora* and *Draparnaldia* (Fig. 14, *B*, *C*), very much branched. The two latter forms are surrounded by masses of transparent jelly, which sometimes reach a length of several centimetres.

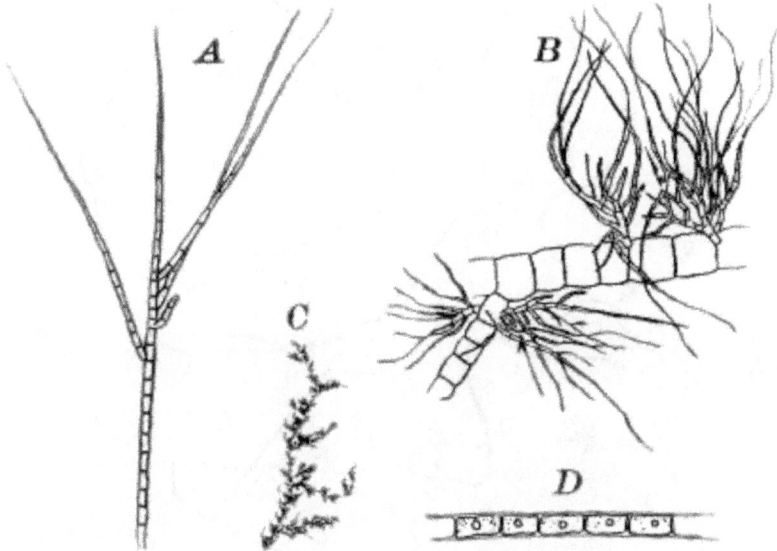

FIG. 14. — *Confervaceæ. A, Stigeoclonium. B, Draparnaldia, × 50. C, a piece of Draparnaldia, × 2. D, part of a filament of Conferva, × 300.*

Among the marine forms related to these may be mentioned the sea lettuce (*Ulva*), shown in Figure 15. The thin, bright-green, leaf-like fronds of this plant are familiar to every seaside student.

FIG. 15.—A plant of sea lettuce (Ulva). One-half natural size.

Somewhat higher than *Cladophora* and its allies, especially in the differentiation of the reproductive parts, are the various species of *Œdogonium* and its relatives. There are numerous species of *Œdogonium* not uncommon in stagnant water growing in company with other algæ, but seldom forming masses by themselves of sufficient size to be recognizable to the naked eye.

The plant is in structure much like *Cladophora*, except that it is unbranched, and the cells have but a single nucleus (Fig. 16, *E*). Even when not fruiting the filaments may usually be recognized by peculiar cap-shaped structures at the top of some of the cells. These arise as the result of certain peculiarities in the process of cell division, which are too complicated to be explained here.

There are two forms of reproduction, non-sexual and sexual. In the first the contents of certain cells escape in the form of large zoöspores (Fig. 16, C), of oval form, having the smaller end colorless and surrounded by a crown of cilia. After a short period of active motion, the zoöspore comes to rest, secretes a cell wall about itself, and the transparent end becomes flattened out into a disc (E, d), by which it fastens itself to some object in the water. The upper part now rapidly elongates, and dividing repeatedly by cross walls, develops into a filament like the original one. In many species special zoöspores are formed, smaller than the ordinary ones, that attach themselves to the filaments bearing the female reproductive organ (oögonium), and grow into small plants bearing the male organ (antheridium), (Fig. 16, B).

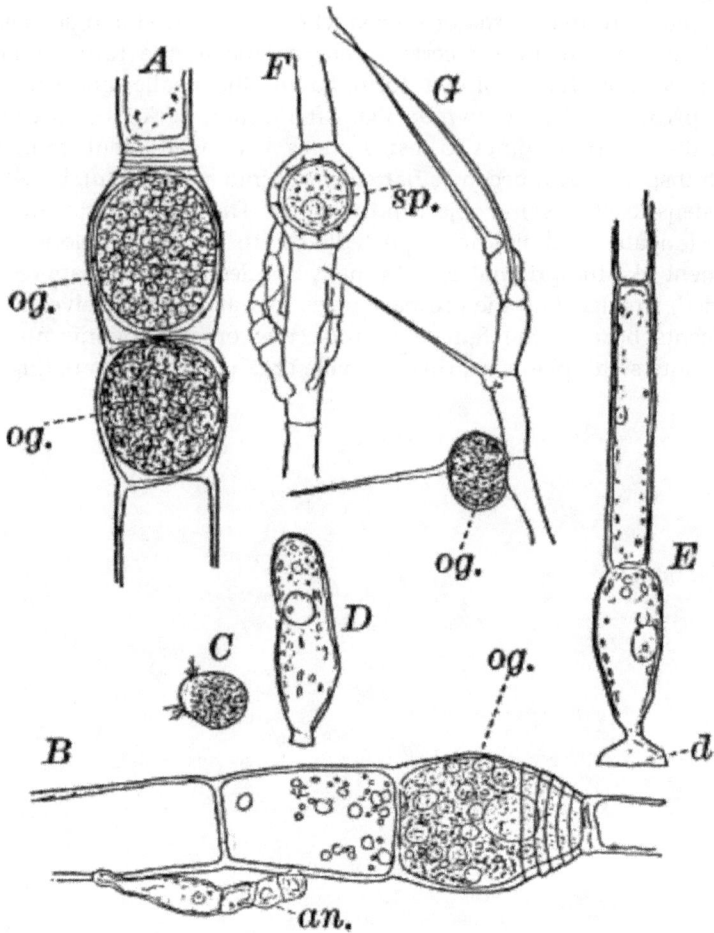

FIG. 16.—A, portion of a filament of Œdogonium, with two oögonia (og.). The lower one shows the opening. B, a similar filament, to which is attached a small male plant with an antheridium (an.). C, a zoöspore of Œdogonium. D, a similar spore germinating. E, base of a filament showing the disc (d) by which it is attached. F, another species of Œdogonium with a ripe spore (sp.). G, part of a plant of Bulbochæte. C, D, × 300; the others × 150.

The sexual reproduction takes place as follows: Certain cells of a filament become distinguished by their denser contents and by an increase in size, becoming oval or nearly globular in form (Fig. 16, A, B). When fully grown, the contents contract and form a naked cell, which sometimes shows a clear area at one point on the surface. This globular mass of protoplasm is the egg cell, or female cell, and the cell containing it is called the "oögonium." When the egg cell is ripe, the oögonium opens by means of a little pore at one side (Fig. 16, A).

In other cells, either of the same filament or else of the small male plants already mentioned, small motile cells, called spermatozoids, are formed. These are much smaller than the egg cell, and resemble the zoöspores in form, but are much smaller, and without chlorophyll. When ripe they are discharged from the cells in which they were formed, and enter the oögonium. By careful observation the student may possibly be able to follow the spermatozoid into the oögonium, where it enters the egg cell at the clear spot on its surface. As a result of the entrance of the spermatozoid (fertilization), the egg cell becomes surrounded by a thick brown wall, and becomes a resting spore. The spore loses its green color, and the wall becomes dark colored and differentiated into several layers, the outer one often provided with spines (Fig. 16, F). As these spores do not germinate for a long time, the process is only known in a comparatively small number of species, and can hardly be followed by the ordinary student.

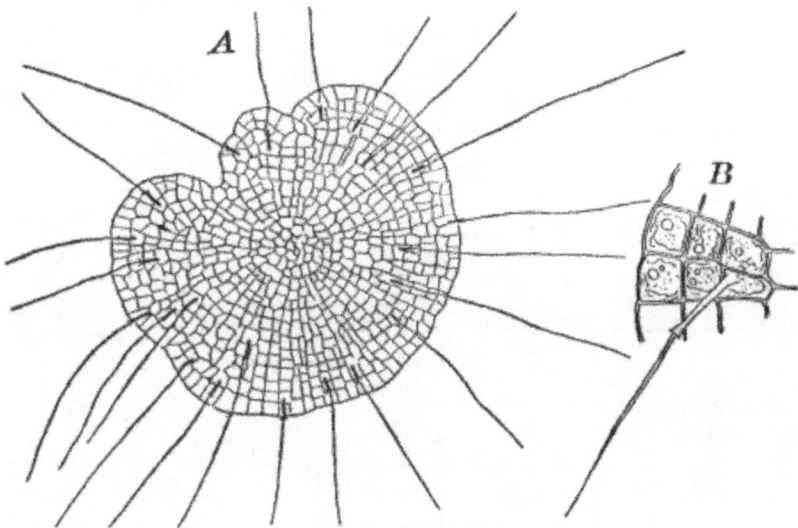

Fig. 17.—A, plant of Coleochæte, × 50. B, a few cells from the margin, with one of the hairs.

Much like *Œdogonium*, but differing in being branched, is the genus *Bulbochæte*, characterized also by hairs swollen at the base, and prolonged into a delicate filament (Fig. 16, *G*).

The highest members of the *Confervaceæ* are those of the genus *Coleochæte* (Fig. 17), of which there are several species found in the United States. These show some striking resemblances to the red seaweeds, and possibly form a transition from the green algæ to the red. The commonest species form bright-green discs, adhering firmly to the stems and floating leaves of water lilies and other aquatics. In aquaria they sometimes attach themselves in large numbers to the glass sides of the vessel.

Growing from the upper surface are numerous hairs, consisting of a short, sheath-like base, including a very long and delicate filament (Fig. 17, *B*). In their methods of reproduction they resemble *Œdogonium*, but the reproductive organs are more specialized.

CHAPTER V.
GREEN ALGÆ — *Continued.*

Order III. — Pond Scums (*CONJUGATÆ*).

THE *Conjugatæ*, while in some respects approaching the *Confervaceæ* in structure, yet differ from them to such an extent in some respects that their close relationship is doubtful. They are very common and familiar plants, some of them forming great floating masses upon the surface of every stagnant pond and ditch, being commonly known as "pond scum." The commonest of these pond scums belong to the genus *Spirogyra*, and one of these will illustrate the characteristics of the order. When in active growth these masses are of a vivid green, and owing to the presence of a gelatinous coating feel slimy, slipping through the hands when one attempts to lift them from the water. Spread out in water, the masses are seen to be composed of slender threads, often many centimetres in length, and showing no sign of branching.

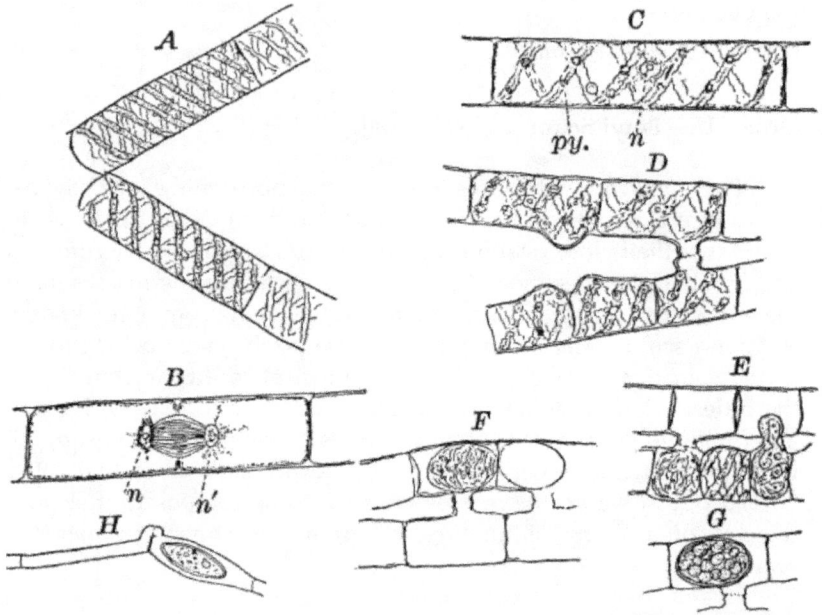

Fig. 18.—A, a filament of a common pond scum (Spirogyra) separating into two parts. B, a cell undergoing division. The cell is seen in optical section, and the chlorophyll bands are omitted, n, n□, the two nuclei. C, a complete cell. n, nucleus. py. pyrenoid. D, E, successive stages in the process of conjugation. G, a ripe spore. H, a form in which conjugation takes place between the cells of the same filament. All × 150.

For microscopical examination the larger species are preferable. When one of these is magnified (Fig. 18, *A, C*), the unbranched filament is shown to be made up of perfectly cylindrical cells, with rather delicate walls. The protoplasm is confined to a thin layer lining the walls, except for numerous fine filaments that radiate from the centrally placed nucleus (*n*), which thus appears suspended in the middle of the cell. The nucleus is large and distinct in the larger species, and has a noticeably large and conspicuous nucleolus. The most noticeable thing about the cell is the green spiral bands running around it. These are the chloroplasts, which in all the *Conjugatæ* are of very peculiar forms. The number of these bands varies much in different species of *Spirogyra*, but is commonly two or three. These chloroplasts, like those of other plants, are not noticeably different in structure from the ordinary protoplasm, as is shown by extracting the chlorophyll, which may be done by placing the plants in alcohol for a short time. This extracts the chlorophyll, but a microscopic examination of the decolored cells shows that the bands remain unchanged, except for the absence of color. These bands are flattened, with irregularly scalloped margins, and at intervals have rounded bodies (pyrenoids) imbedded in them (Fig. 18, *C, py.*). The pyrenoids, especially when the plant has been exposed to the light for some time, are surrounded by a circle of small granules, which become bluish when iodine is applied, showing them to be starch. (To show the effect of iodine on starch on a large scale, mix a little flour, which is nearly all starch, with water, and add a little iodine. The starch will immediately become colored blue, varying in intensity with the amount of iodine.) The cells divide much as in *Cladophora*, but the nucleus here takes part in the process. The division naturally occurs only at night, but by reducing the temperature at night to near the freezing point (4° C., or a little lower), the process may be checked. The experiment is most conveniently made when the temperature out of doors approaches the freezing point. Then it is only necessary to keep the plants in a warm room until about 10 P.M., when they may be put out of doors for the night. On bringing them in in the morning, the division will begin almost at once, and may be easily studied. The nucleus divides into two parts, which remain for a time connected by delicate threads (Fig. 18, *B*), that finally disappear. At first no nucleoli are present in the daughter nuclei, but they appear before the division is complete.

New filaments are formed by the breaking up of the old ones, this sometimes being very rapid. As the cells break apart, the free ends bulge strongly, showing the pressure exerted upon the cell wall by the contents (Fig. 18, *A*).

Spores like those of *Œdogonium* are formed, but the process is somewhat different. It occurs in most species late in the spring, but may sometimes be met with at other times. The masses of fruiting plants usually appear brownish colored. If spores have been formed they can, in the larger species at least, be seen with a hand lens, appearing as rows of dark-colored specks.

Two filaments lying side by side send out protuberances of the cell wall that grow toward each other until they touch (Fig. 18, *D*). At the point of contact, the wall is absorbed, forming a continuous channel from one cell to the other. This process usually takes place in all the cells of the two filaments, so that the two filaments, connected by tubes at regular intervals, have the form of a ladder.

In some species adjoining cells of the same filament become connected, the tubes being formed at the end of the cells (Fig. 18, *H*), and the cell in which the spore is formed enlarges.

Soon after the channel is completed, the contents of one cell flow slowly through it into the neighboring cell, and the protoplasm of the two fuses into one mass. (The union of the nuclei has also been observed.) The young spore thus formed contracts somewhat, becoming oval in form, and soon secretes a thick wall, colorless at first, but afterwards becoming brown and more or less opaque. The chlorophyll bands, although much crowded, are at first distinguishable, but later lose the chlorophyll, and become unrecognizable. Like the resting spores of *Œdogonium* these require a long period of rest before germinating.

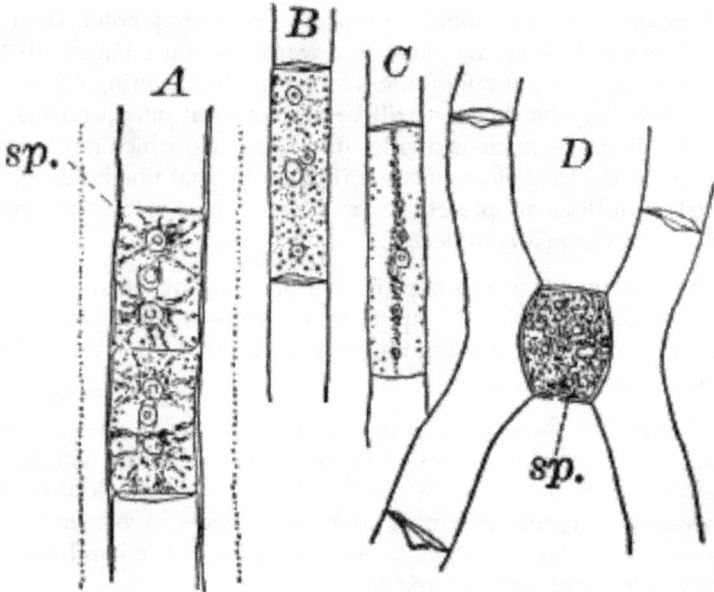

FIG. 19.—*Forms of Zygnemaceæ. A, Zygnema. B, C, D, Mesocarpus. All × 150.*

There are various genera of the pond scums, differing in the form of the chloroplasts and also in the position of the spores. Of these may be mentioned *Zygnema* (Fig. 19, *A*), with two star-shaped chloroplasts in each cell, and *Mesocarpus* (Fig. 19, *B, D*), in which the single chloroplast has the form of a thin median plate. (B shows the appearance from in front, C from the side, showing the thickness of the plate.) *Mesocarpus* and the allied genera have the spore formed between the filaments, the contents of both the uniting cells leaving them.

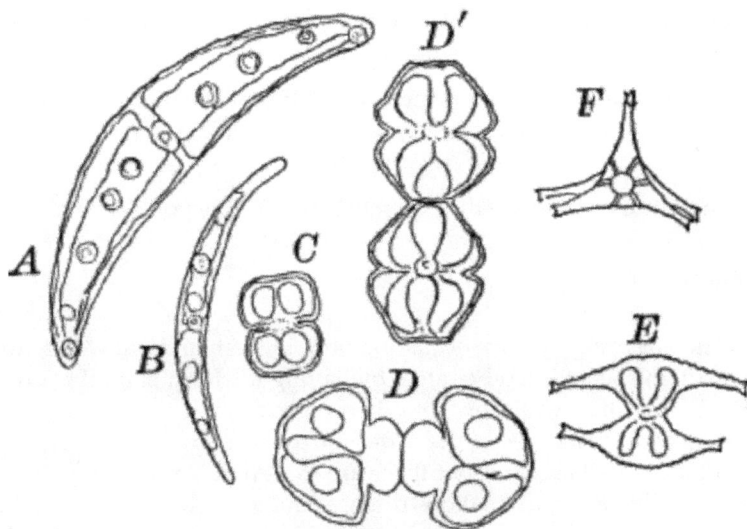

FIG. 20.—*Forms of Desmids. A, B, Closterium. C, D, D', Cosmarium. D, and D' show the process of division. E, F, Staurastrum; E seen from the side, F from the end.*

Evidently related to the pond scums, but differing in being for the most part strictly unicellular, are the desmids (Fig. 20). They are confined to fresh water, and seldom occur in masses of sufficient size to be seen with the naked eye, usually being found associated with pond scums or other filamentous forms. Many of the most beautiful forms may be obtained by examining the matter adhering to the leaves and stems of many floating water plants, especially the bladder weed (*Utricularia*) and other fine-leaved aquatics.

The desmids include the most beautiful examples of unicellular plants to be met with, the cells having extremely elegant outlines. The cell shows a division into two parts, and is often constricted in the middle, each division having a single large chloroplast of peculiar form. The central part of the cell in which the nucleus lies is colorless.

40

Among the commonest forms, often growing with *Spirogyra*, are various species of *Closterium* (Fig. 20, *A*, *B*), recognizable at once by their crescent shape. The cell appears bright green, except at the ends and in the middle. The large chloroplast in each half is composed of six longitudinal plates, united at the axis of the cell. Several large pyrenoids are always found, often forming a regular line through the central axis. At each end of the cell is a vacuole containing small granules that show an active dancing movement.

The desmids often have the power of movement, swimming or creeping slowly over the slide as we examine them, but the mechanism of these movements is still doubtful.

In their reproduction they closely resemble the pond scums.

Order IV. — *SIPHONEÆ*.

The *Siphoneæ* are algæ occurring both in fresh and salt water, and are distinguished from other algæ by having the form of a tube, undivided by partition walls, except when reproduction occurs. The only common representatives of the order in fresh water are those belonging to the genus *Vaucheria*, but these are to be had almost everywhere. They usually occur in shallow ditches and ponds, growing on the bottom, or not infrequently becoming free, and floating where the water is deeper. They form large, dark green, felted masses, and are sometimes known as "green felts." Some species grow also on the wet ground about springs. An examination of one of the masses shows it to be made up of closely matted, hair-like threads, each of which is an individual plant.

In transferring the plants to the slide for microscopic examination, they must be handled very carefully, as they are very easily injured. Each thread is a long tube, branching sometimes, but not divided into cells as in *Spirogyra* or *Cladophora*. If we follow it to the tip, the contents here will be found to be denser, this being the growing point. By careful focusing it is easy to show that the protoplasm is confined to a thin layer lining the wall, the central cavity of the tube being filled with cell sap. In the protoplasm are numerous elongated chloroplasts (*cl.*). and a larger or smaller number of small, shining, globular bodies (*ol.*). These latter are drops of oil, and, when the filaments are injured, sometimes run together, and form drops of large size. No nucleus can be seen in the living plant, but by treatment with

41

chromic acid and staining, numerous very small nuclei may be demonstrated.

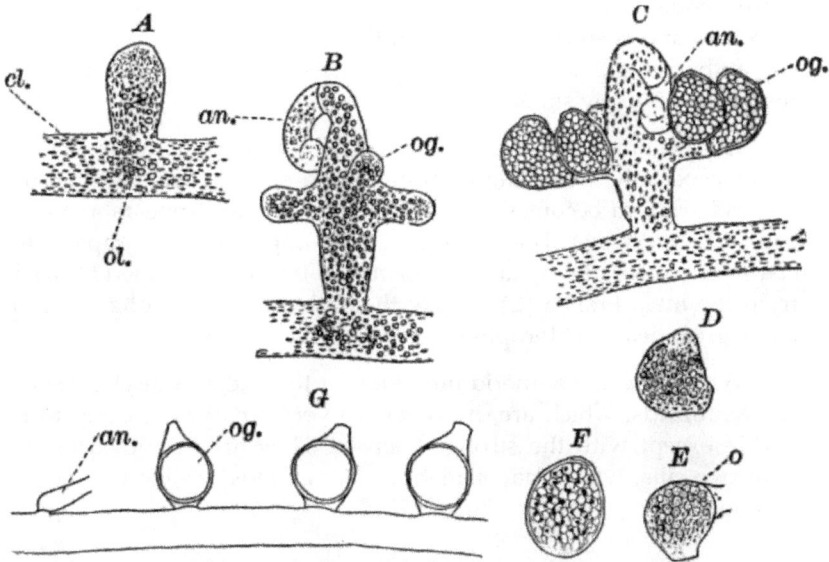

FIG. 21.—A, C, successive stages in the development of the sexual organs of a green felt (Vaucheria). an. antheridium. og. oögonium. D, a ripe oögonium. E, the same after it has opened. o, the egg cell. F, a ripe spore. G, a species in which the sexual organs are borne separately on the main filament. A, F, × 150. G, × 50. cl. chloroplasts. ol. oil.

When the filaments are growing upon the ground, or at the bottom of shallow water, the lower end is colorless, and forms a more or less branching root-like structure, fastening it to the earth. These rootlets, like the rest of the filament, are undivided by walls.

One of the commonest and at the same time most characteristic species is *Vaucheria racemosa* (Fig. 21, A, F). The plant multiplies nonsexually by branches pinched off by a constriction at the point where they join the main filament, or by the filament itself becoming constricted and separating into several parts, each one constituting a new individual.

The sexual organs are formed on special branches, and their arrangement is such as to make the species instantly recognizable.

42

The first sign of their development is the formation of a short branch (Fig. 21, *A*) growing out at right angles to the main filament. This branch becomes club-shaped, and the end somewhat pointed and more slender, and curves over. This slender, curved portion is almost colorless, and is soon shut off from the rest of the branch. It is called an "antheridium," and within are produced, by internal division, numerous excessively small spermatozoids.

As the branch grows, its contents become very dense, the oil drops especially increasing in number and size. About the time that the antheridium becomes shut off, a circle of buds appears about its base (Fig. 21, *B, og.*). These are the young oögonia, which rapidly increase in size, assuming an oval form, and become separated by walls from the main branch (*C*). Unlike the antheridium, the oögonia contain a great deal of chlorophyll, appearing deep green.

When ripe, the antheridium opens at the end and discharges the spermatozoids, which are, however, so very small as scarcely to be visible except with the strongest lenses. They are little oval bodies with two cilia, which may sometimes be rendered visible by staining with iodine.

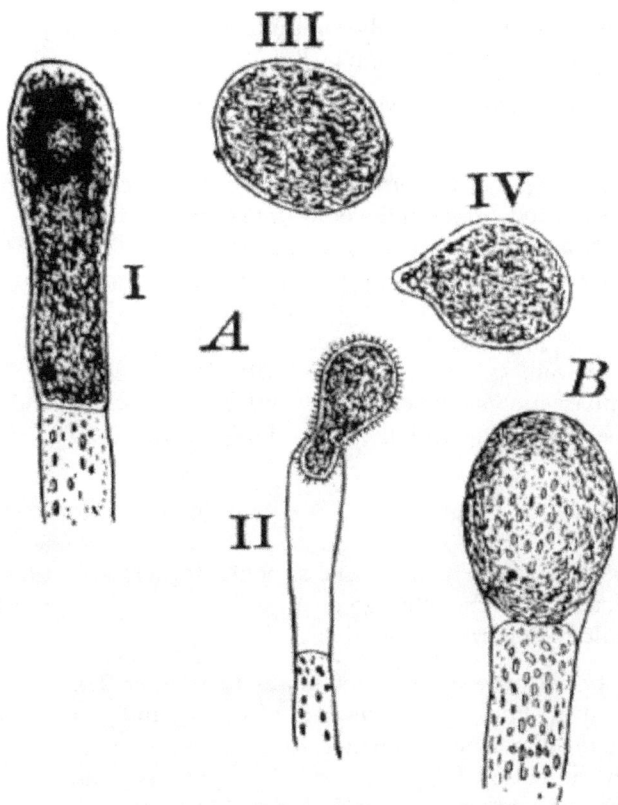

FIG. 22.—*A, non-sexual reproduction in Vaucheria sessilis. B, non-sexual spore of V. geminata,* × 50.

The oögonia, which at first are uniformly colored, just before maturity show a colorless space at the top, from which the chloroplasts and oil drops have disappeared (*D*), and at the same time this portion pushes out in the form of a short beak. Soon after the wall is absorbed at this point, and a portion of the contents is forced out, leaving an opening, and at the same time the remaining contents contract to form a round mass, the germ or egg cell (Fig. 21, *E, o*). Almost as soon as the oögonium opens, the spermatozoids collect about it and enter; but, on account of their minuteness, it is almost impossible to follow them into the egg cell, or to determine whether several or only one enter. The fertilized egg cell becomes almost at once surrounded by a wall, which rapidly thickens, and forms a resting spore. As the spore ripens, it loses its green color, becoming colorless, with a few reddish brown specks scattered through it (*F*).

In some species the sexual organs are borne directly on the filament (Fig. 21, *G*).

Large zoöspores are formed in some of the green felts (Fig. 22, *A*), and are produced singly in the ends of branches that become swollen, dark green, and filled with very dense protoplasm. This end becomes separated by a wall from the rest of the branch, the end opens, and the contents escape as a very large zoöspore, covered with numerous short cilia (*A* II). After a short period of activity, this loses its cilia, develops a wall, and begins to grow (III, IV). Other species (*B*) produce similar spores, which, however, are not motile, and remain within the mother cell until they are set free by the decay of its wall.

Order V. — *CHARACEÆ*.

The *Characeæ*, or stone-worts, as some of them are called, are so very different from the other green algæ that it is highly probable that they should be separated from them.

The type of the order is the genus *Chara* (Fig. 23), called stone-worts from the coating of carbonate of lime found in most of them, giving them a harsh, stony texture. Several species are common growing upon the bottom of ponds and slow streams, and range in size from a few centimetres to a metre or more in height.

The plant (Fig. 23, *A*) consists of a central jointed axis with circles of leaves at each joint or node. The distance between the nodes (internodes) may in the larger species reach a length of several centimetres. The leaves are slender, cylindrical structures, and like the stem divided into nodes and internodes, and have at the nodes delicate leaflets.

At each joint of the leaf, in fruiting specimens, attached to the inner side, are borne two small, roundish bodies, in the commoner species of a reddish color (Fig. 23, *A*, *r*). The lower of the two is globular, and bright scarlet in color; the other, more oval and duller.

Examined with a lens the main axis presents a striated appearance. The whole plant is harsh to the touch and brittle, owing to the limy coating. It is fastened to the ground by fine, colorless hairs, or rootlets.

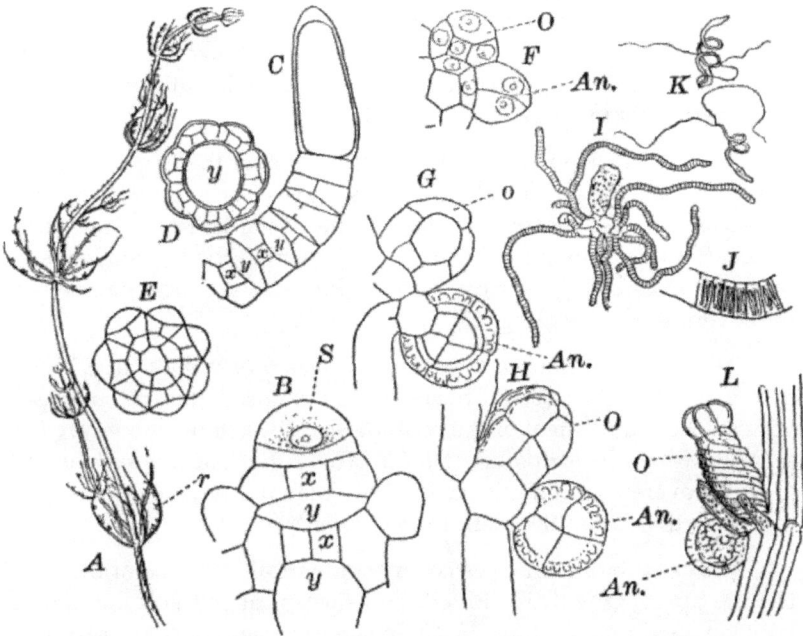

Fig. 23.—A, plant of a stone-wort (Chara), one-half natural size. r, reproductive organs. B, longitudinal section through the apex. S, apical cell. x, nodes. y, internodes. C, a young leaf. D, cross section of an internode. E, of a node of a somewhat older leaf. F, G, young sexual organs seen in optical section. o, oögonium. An. antheridium. H, superficial view. G, I, group of filaments containing spermatozoids. J, a small portion of one of these more magnified, showing a spermatozoid in each cell. K, free spermatozoids. L, a piece of a leaf with ripe oögonium (o), and antheridium (An.). B, H, × 150. J, K, × 300. I, × 50. L, × 25.

By making a series of longitudinal sections with a sharp razor through the top of the plant, and magnifying sufficiently, it is found to end in a single, nearly hemispherical cell (Fig. 23, *B, S*). This from its position is called the "apical cell," and from it are derived all the tissues of the plant. Segments are cut off from its base, and these divide again into two by a wall parallel to the first. Of the two cells thus formed one undergoes no further division and forms the central cell of

46

an internode (*y*); the other divides repeatedly, forming a node or joint (*x*).

As the arrangement of these cells is essentially the same in the leaves and stem, we will examine it in the former, as by cutting several cross-sections of the whole bunch of young leaves near the top of the plant, we shall pretty certainly get some sections through a joint. The arrangement is shown in Figure 23, *E*.

As the stem grows, a covering is formed over the large internodal cell (*y*) by the growth of cells from the nodes. These grow both from above and below, meeting in the middle of the internode and completely hiding the long axial cell. A section across the internode shows the large axial cell (*y*) surrounded by the regularly arranged cells of the covering or cortex (Fig. 23, *D*).

All the cells contain a layer of protoplasm next the wall with numerous oval chloroplasts. If the cells are uninjured, they often show a very marked movement of the protoplasm. These movements are best seen, however, in forms like *Nitella*, where the long internodal cells are not covered with a cortex. In *Chara* they are most evident in the root hairs that fasten the plant to the ground.

The growth of the leaves is almost identical with that of the stem, but the apical growth is limited, and the apical cell becomes finally very long and pointed (Fig. 23, *C*). In some species the chloroplasts are reddish in the young cells, assuming their green color as the cells approach maturity.

The plant multiplies non-sexually by means of special branches that may become detached, but there are no non-sexual spores formed.

The sexual organs have already been noticed arising in pairs at the joints of the leaves. The oögonium is formed above, the antheridium below.

The young oögonium (*F, O*) consists of a central cell, below which is a smaller one surrounded by a circle of five others, which do not at first project above the central cell, but later completely envelop it (*G*). Each of these five cells early becomes divided into an upper and a lower one, the latter becoming twisted as it elongates, and the central cell later has a small cell cut off from its base by an oblique wall. The central cell forms the egg cell, which in the ripe oögonium (*L, O*) is surrounded by five, spirally twisted cells, and crowned by a circle of five smaller ones, which become of a yellowish color when full grown. They separate at the time of fertilization to allow the spermatozoids to enter the oögonium.

The antheridium consists at first of a basal cell and a terminal one. The latter, which is nearly globular, divides into eight nearly similar cells by walls passing through the centre. In each of these eight cells two walls are next formed parallel to the outer surface, so that the antheridium (apart from the basal cell) contains twenty-four cells arranged in three concentric series (G, an.). These cells, especially the outer ones, develop a great amount of a red pigment, giving the antheridium its characteristic color.

The diameter of the antheridium now increases rapidly, and the central cells separate, leaving a large space within. Of the inner cells, the second series, while not increasing in diameter, elongate, assuming an oblong form, and from the innermost are developed long filaments (I, J) composed of a single row of cells, in each of which is formed a spermatozoid.

The eight outer cells are nearly triangular in outline, fitting together by deeply indented margins, and having the oblong cells with the attached filaments upon their inner faces.

If a ripe antheridium is crushed in a drop of water, after lying a few minutes the spermatozoids will escape through small openings in the side of the cells. They are much larger than any we have met with. Each is a colorless, spiral thread with about three coils, one end being somewhat dilated with a few granules; the other more pointed, and bearing two extremely long and delicate cilia (K). To see the cilia it is necessary to kill the spermatozoids with iodine or some other reagent.

After fertilization the outer cells of the oögonium become very hard, and the whole falls off, germinating after a sufficient period of rest.

According to the accounts of Pringsheim and others, the young plant consists at first of a row of elongated cells, upon which a bud is formed that develops into the perfect plant.

There are two families of the Characeæ, the Chareæ, of which Chara is the type, and the Nitelleæ, represented by various species of Nitella and Tolypella. The second family have the internodes without any cortex—that is, consisting of a single long cell; and the crown at the top of the oögonium is composed of ten cells instead of five. They are also destitute of the limy coating of the *Chareæ*.

Both as regards the structure of the plant itself, as well as the reproductive organs, especially the very complex antheridium, the *Characeæ* are very widely separated from any other group of plants, either above or below them.

CHAPTER VI.
THE BROWN ALGÆ (*Phæophyceæ*).

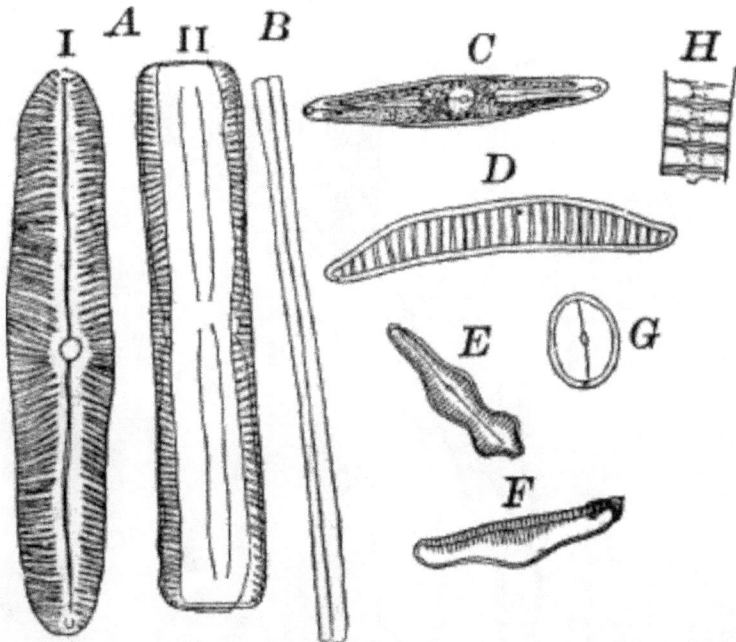

FIG. 24.—*Forms of diatoms. A, Pinnularia. I, seen from above; II, from the side. B, Fragillaria (?). C, Navicula. D, F, Eunotia. E, Gomphonema. G, Cocconeis. H, Diatoma. All × 300.*

THESE plants are all characterized by the presence of a brown pigment, in addition to the chlorophyll, which almost entirely conceals the latter, giving the plants a brownish color, ranging from a light yellowish brown to nearly black. One order of plants that possibly belongs here (*Diatomaceæ*) are single celled, but the others are for the most part large seaweeds. The diatoms, which are placed in this class simply on account of the color, are probably not closely related to the other brown algæ, but just where they should be placed is difficult to say. In some respects they approach quite closely the desmids, and are not infrequently regarded as related to them. They are among the commonest of organisms occurring everywhere in stagnant and running water, both fresh and salt, forming usually, slimy, yellowish coatings on stones, mud, aquatic plants, etc. Like the desmids they may be single or united into filaments, and not infrequently are attached by means of a delicate gelatinous stalk (Fig. 25).

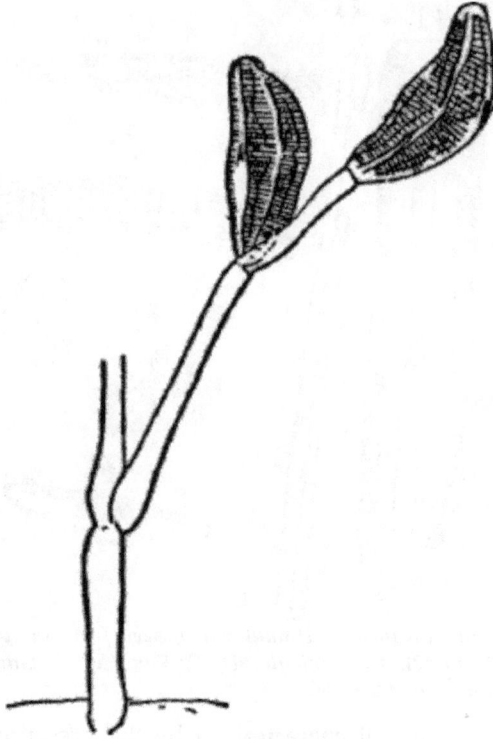

FIG. 25. — *Diatoms attached by a gelatinous stalk.* × 150

They are at once distinguished from the desmids by their color, which is always some shade of yellowish or reddish brown. The commonest forms, *e.g. Navicula* (Fig. 24, C), are boat-shaped when seen from above, but there is great variety in this respect. The cell wall is always impregnated with large amounts of flint, so that after the cell dies its shape is perfectly preserved, the flint making a perfect cast of it, looking like glass. These flinty shells exhibit wonderfully beautiful and delicate markings which are sometimes so fine as to test the best lenses to make them out.

This shell is composed of two parts, one shutting over the other like a pill box and its cover. This arrangement is best seen in such large forms as *Pinnularia* (Fig. 24, A II).

Most of the diatoms show movements, swimming slowly or gliding over solid substances; but like the movements of *Oscillaria* and the desmids, the movements are not satisfactorily understood, although several explanations have been offered.

They resemble somewhat the desmids in their reproduction.

The True Brown Algæ.

These are all marine forms, many of great size, reaching a length in some cases of a hundred metres or more, and showing a good deal of differentiation in their tissues and organs.

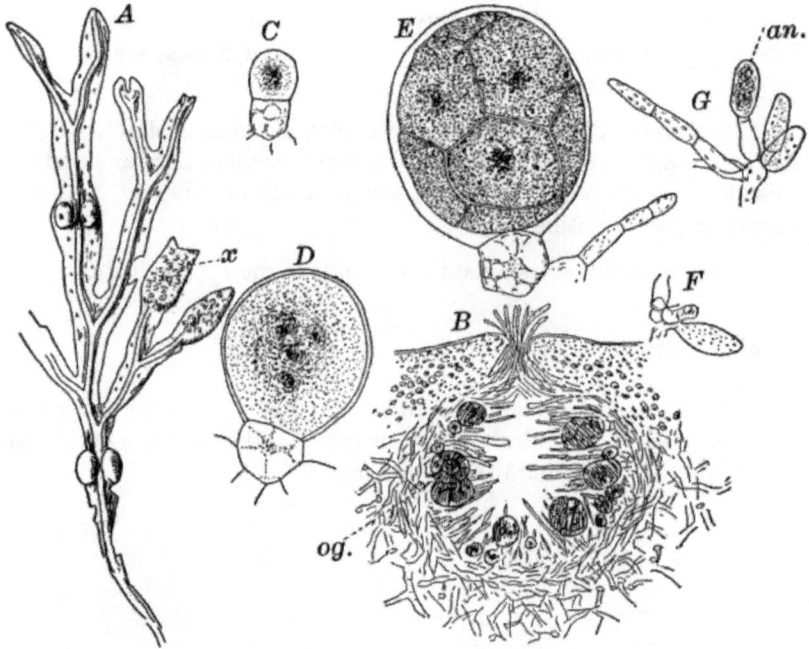

Fig. 26.—A, a branch of common rock weed (Fucus), one-half natural size. x, end of a branch bearing conceptacles. B, section through a conceptacle containing oögonia (og.), × 25. C, E, successive stages in the development of the oögonium, × 150. F, G, antheridia. In G, one of the antheridia has discharged the mass of spermatozoids (an.), × 150.

One of the commonest forms is the ordinary rock weed (Fucus), which covers the rocks of our northeastern coast with a heavy drapery for several feet above low-water mark, so that the plants are completely exposed as the tide recedes. The commonest species, F. vesiculosus (Fig. 26, A), is distinguished by the air sacs with which the stems are provided. The plant is attached to the rock by means of a sort of disc or root from which springs a stem of tough, leathery texture, and forking regularly at intervals, so that the ultimate branches are very numerous, and the plant may reach a length of a metre or more. The branches are flattened and leaf-like, the centre traversed by a thickened midrib. The end of the growing branches is occupied by a transversely elongated pit or depression. The growing point is at the bottom of this pit, and by a regular forking of the growing point the symmetrical branching of the plant is brought about. Scattered over the surface are little circular pits through whose openings protrude bunches of fine hairs. When wet the plant is flexible and leathery, but it may become quite dry and hard without suffering, as may be seen when the plants are exposed to the sun at low tide.

The air bladders are placed in pairs, for the most part, and buoy up the plant, bringing it up to the surface when covered with water.

The interior of the plant is very soft and gelatinous, while the outer part forms a sort of tough rind of much firmer consistence. The ends of some of the branches (Fig. 26, *A*, *x*) are usually much swollen, and the surface covered with little elevations from which may often be seen protruding clusters of hairs like those arising from the other parts of the plant. A section through one of these enlarged ends shows that each elevation corresponds to a cavity situated below it. On some of the plants these cavities are filled with an orange-yellow mass; in others there are a number of roundish olive-brown bodies large enough to be easily seen. The yellow masses are masses of antheridia; the round bodies, the oögonia.

If the plants are gathered while wet, and packed so as to prevent evaporation of the water, they will keep perfectly for several days, and may readily be shipped for long distances. If they are to be studied away from the seashore, sections for microscopic examination should be mounted in salt water (about 3 parts in weight of common salt to 100 of water). If fresh material is not to be had, dried specimens or alcoholic material will answer pretty well.

To study the minute structure of the plant, make a thin cross-section, and mount in salt water. The inner part or pith is composed of loosely arranged, elongated cells, placed end to end, and forming an irregular network, the large spaces between filled with the mucilaginous substance derived from the altered outer walls of these cells. This mucilage is hard when dry, but swells up enormously in water, especially fresh water. The cells grow smaller and more compact toward the outside of the section, until there are no spaces of any size between those of the outside or rind. The cells contain small chloroplasts like those of the higher plants, but owing to the presence of the brown pigment found in all of the class, in addition to the chlorophyll, they appear golden brown instead of green.

No non-sexual reproductive bodies are known in the rock weeds, beyond small branches that occur in clusters on the margins of the main branches, and probably become detached, forming new plants.

In some of the lower forms, however, *e.g. Ectocarpus* and *Laminaria* (Fig. 28, *A, C*), zoöspores are formed.

The sexual organs of the rock weed, as we have already seen, are borne in special cavities (conceptacles) in the enlarged ends of some of the branches. In the species here figured, *F. vesiculosus*, the antheridia and oögonia are borne on separate plants; but in others, *e.g. F. platycarpus*, they are both in the same conceptacle.

The walls of the conceptacle (Fig. 26, *B*) are composed of closely interwoven filaments, from which grow inward numerous hairs, filling up the space within, and often extending out through the opening at the top.

The reproductive bodies arise from the base of these hairs. The oögonia (Fig. 26, *C, E*) arise as nearly colorless cells, that early become divided into two cells, a short basal cell or stalk and a larger terminal one, the oögonium proper. The latter enlarges rapidly, and its contents divide into eight parts. The division is at first indicated by a division of the central portion, which includes the nucleus, and is colored brown, into two, four, and finally eight parts, after which walls are formed between these. The brown color spreads until the whole oögonium is of a nearly uniform olive-brown tint.

When ripe, the upper part of the oögonium dissolves, allowing the eight cells, still enclosed in a delicate membrane, to escape (Fig. 27, *H*). Finally, the walls separating the inner cells of the oögonium become also absorbed, as well as the surrounding membrane, and the eight egg cells escape into the water (Fig. 27, *I*) as naked balls of protoplasm, in which a central nucleus may be dimly seen.

The antheridia (Fig. 26, *F, G*) are small oblong cells, at first colorless, but when ripe containing numerous glistening, reddish brown dots, each of which is part of a spermatozoid. When ripe, the contents of the antheridium are forced out into the water (*G*), leaving the empty outer wall behind, but still surrounded by a thin membrane. After a few minutes this membrane is dissolved, and the spermatozoids are set free. These (Fig. 27, *K*) are oval in form, with two long cilia attached to the side where the brown speck, seen while still within the antheridium, is conspicuous.

The act of fertilization may be easily observed by laying fresh antheridia into a drop of water containing recently discharged egg cells. To obtain these, all that is necessary is to allow freshly gathered plants to remain in the air until they are somewhat dry, when the ripe sexual cells will be discharged from the openings of the conceptacles, exuding as little drops, those with antheridia being orange-yellow; the masses of oögonia, olive. Within a few minutes after putting the oögonia into water, the egg cells may be seen to escape into the water, when some of the antheridia may be added. The spermatozoids will be quickly discharged, and collect immediately in great numbers about the egg cells, to which they apply themselves closely, often setting them in rotation by the movements of their cilia, and presenting a most extraordinary spectacle (*J*). Owing to the small size of the spermatozoids, and the opacity of the eggs, it is impossible to see whether more than one spermatozoid penetrates it; but from what is known in other cases it is not likely. The egg now secretes a wall about itself, and within a short time begins to grow. It becomes pear-shaped, the narrow portion becoming attached to the parent plant or to some other object by means of rootlets, and the upper part grows into the body of the young plant (Fig. 27, *M*).

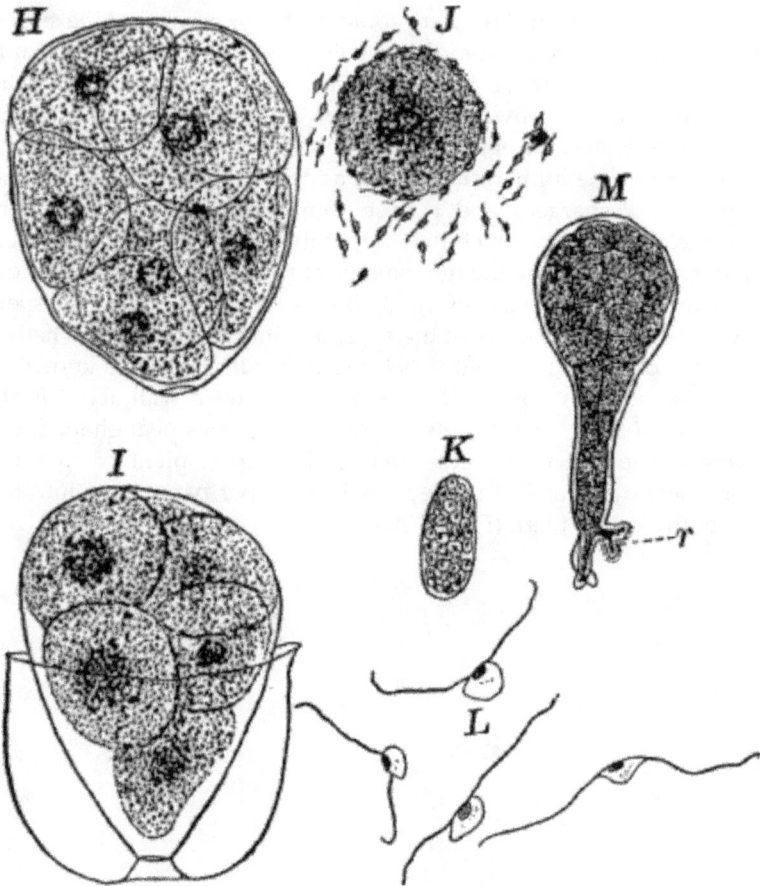

FIG. 27.—H, *the eight egg cells still surrounded by the inner membrane of the oögonium. I, the egg cells escaping into the water. J, a single egg cell surrounded by spermatozoids. K, mass of spermatozoids surrounded by the inner membrane of the antheridium. L, spermatozoids. M, young plant. r, the roots. K, × 300; L, × 600; the others, × 150.*

The simpler brown seaweeds, so far as known, multiply only by means of zoöspores, which may grow directly into new plants, or, as has been observed in some species, two zoöspores will first unite. A few, like *Ectocarpus* (Fig. 28, *A*), are simple, branched filaments, but most are large plants with complex tissues. Of the latter, a familiar example is the common kelp, "devil's apron" (*Laminaria*), often three to four metres in length, with a stout stalk, provided with root-like organs, by which it is firmly fastened. Above, it expands into a broad, leaf-like frond, which in some species is divided into strips. Related to the kelps is the giant kelp of the Pacific (*Macrocystis*), which is said sometimes to reach a length of three hundred metres.

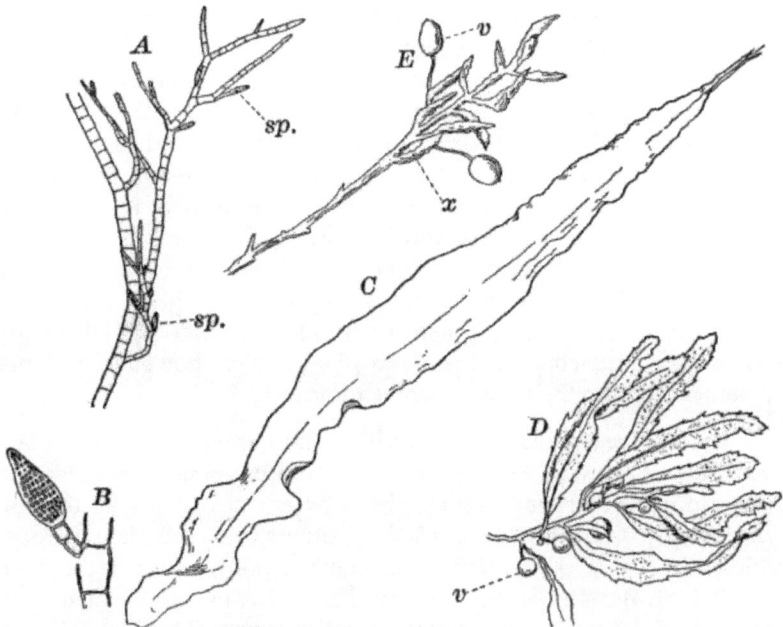

FIG. 28. — *Forms of brown seaweeds. A, Ectocarpus, × 50. Sporangia (sp.). B, a single sporangium, × 150. C, kelp (Laminaria), × ⅛. D, E, gulf weed (Sargassum). D, one-half natural size. E, natural size. v, air bladders. x, conceptacle bearing branches.*

The highest of the class are the gulf weeds (*Sargassum*), plants of the warmer seas, but one species of which is found from Cape Cod southward (Fig. 28, *D, E*). These plants possess distinct stems and leaves, and there are stalked air bladders, looking like berries, giving the plant a striking resemblance to the higher land plants.

CHAPTER VII.

Class III.—The Red Algæ (Rhodophyceæ).

THESE are among the most beautiful and interesting members of the plant kingdom, both on account of their beautiful colors and the exquisitely graceful forms exhibited by many of them. Unfortunately for inland students they are, with few exceptions, confined to salt water, and consequently fresh material is not available. Nevertheless, enough can be done with dried material to get a good idea of their general appearance, and the fruiting plants can be readily preserved in strong alcohol. Specimens, simply dried, may be kept for an indefinite period, and on being placed in water will assume perfectly the appearance of the living plants. Prolonged exposure, however, to the action of fresh water extracts the red pigment that gives them their characteristic color. This pigment is found in the chlorophyll bodies, and usually quite conceals the chlorophyll, which, however, becomes evident so soon as the red pigment is removed.

The red seaweeds differ much in the complexity of the plant body, but all agree in the presence of the red pigment, and, at least in the main, in their reproduction. The simpler ones consist of rows of cells, usually branching like *Cladophora*; others form cell plates comparable to *Ulva* (Fig. 30, *C, D*); while others, among which is the well-known Irish moss (*Chondrus*), form plants of considerable size, with pretty well differentiated tissues. In such forms the outer cells are smaller and firmer, constituting a sort of rind; while the inner portions are made up of larger and looser cells, and may be called the pith. Between these extremes are all intermediate forms.

They usually grow attached to rocks, shells, wood, or other plants, such as the kelps and even the larger red seaweeds. They are most abundant in the warmer seas, but still a considerable number may be found in all parts of the ocean, even extending into the Arctic regions.

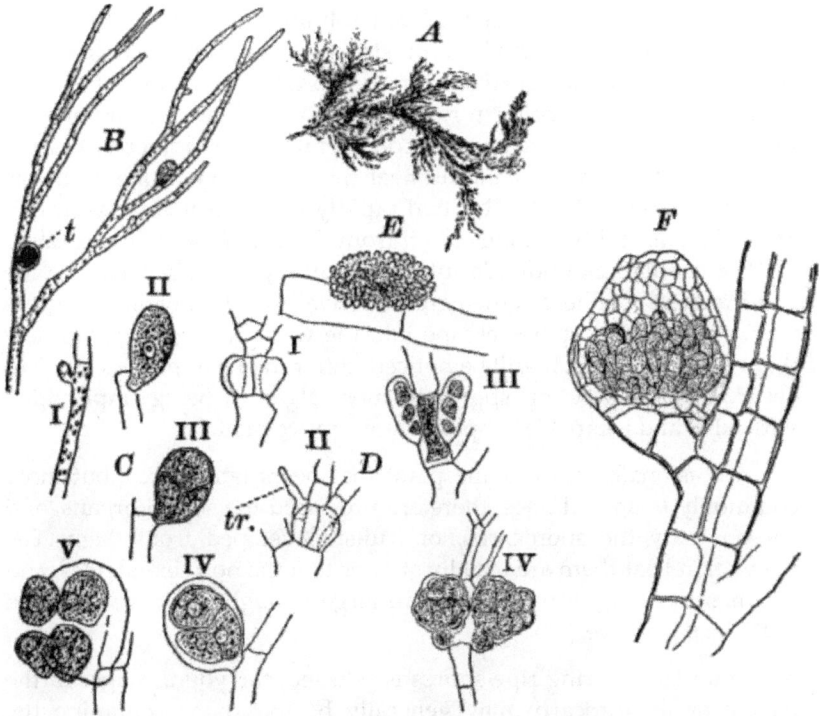

FIG. 29.—*A, a red seaweed (Callithamnion), of the natural size. B, a piece of the same, × 50. t, tetraspores. C I–V, successive stages in the development of the tetraspores, × 150. D I, II young procarps. tr. trichogyne. III, young; IV, ripe spore fruit. I, III, × 150. IV, × 50. E, an antheridium, × 150. F, spore fruit of Polysiphonia. The spores are here surrounded by a case, × 50.*

The methods of reproduction may be best illustrated by a specific example, and preferably one of the simpler ones, as these are most readily studied microscopically.

The form here illustrated (*Callithamnion*) grows attached to wharves, etc., below low-water mark, and is extremely delicate, collapsing completely when removed from the water. The color is a bright rosy red, and with its graceful form and extreme delicacy it makes one of the most beautiful of the group.

If alcoholic material is used, it may be mounted for examination either in water or very dilute glycerine.

The plant is composed of much-branched, slender filaments, closely resembling *Cladophora* in structure, but with smaller cells (Fig. 29, *B*). The non-sexual reproduction is by means of special spores, which from being formed in groups of four, are known as tetraspores. In the species under consideration the mother cell of the tetraspores arises as a small bud near the upper end of one of the ordinary cells (Fig. 29, *C* I). This bud rapidly increases in size, assuming an oval form, and becoming cut off from the cell of the stem (Fig. 29, *C* II). The contents now divide into four equal parts, arranged like the quadrants of a sphere. When ripe, the wall of the mother cell gives way, and the four spores escape into the water and give rise to new plants. These spores, it will be noticed, differ in one important particular from corresponding spores in most algæ, in being unprovided with cilia, and incapable of spontaneous movement.

Occasionally in the same plant that bears tetraspores, but more commonly in special ones, there are produced the sexual organs, and subsequently the sporocarps, or fruits, developed from them. The plants that bear them are usually stouter that the non-sexual ones, and the masses of ripe carpospores are large enough to be readily seen with the naked eye.

If a plant bearing ripe spores is selected, the young stages of the female organ (procarp) may generally be found by examining the younger parts of the plant. The procarp arises from a single cell of the filament. This cell undergoes division by a series of longitudinal walls into a central cell and about four peripheral ones (Fig. 29, *D* I). One of the latter divides next into an upper and a lower cell, the former growing out into a long, colorless appendage known as a trichogyne (Fig. 29, *D, tr.*).

The antheridia (Fig. 29, *E*) are hemispherical masses of closely set colorless cells, each of which develops a single spermatozoid which, like the tetraspores, is destitute of cilia, and is dependent upon the movement of the water to convey it to the neighborhood of the procarp. Occasionally one of these spermatozoids may be found attached to the trichogyne, and in this way fertilization is effected. Curiously enough, neither the cell which is immediately fertilized, nor the one beneath it, undergo any further change; but two of the other peripheral cells on opposite sides of the filament grow rapidly and develop into large, irregular masses of spores (Fig. 29, *D* III, IV).

While the plant here described may be taken as a type of the group, it must be borne in mind that many of them differ widely, not only in the structure of the plant body, but in the complexity of the sexual organs and spores as well. The tetraspores are often imbedded in the tissues of the plant, or may be in special receptacles, nor are they always arranged in the same way as here described, and the same is true of the carpospores. These latter are in some of the higher forms, *e.g. Polysiphonia* (Fig. 29, *F*), contained in urn-shaped receptacles, or they may be buried within the tissues of the plant.

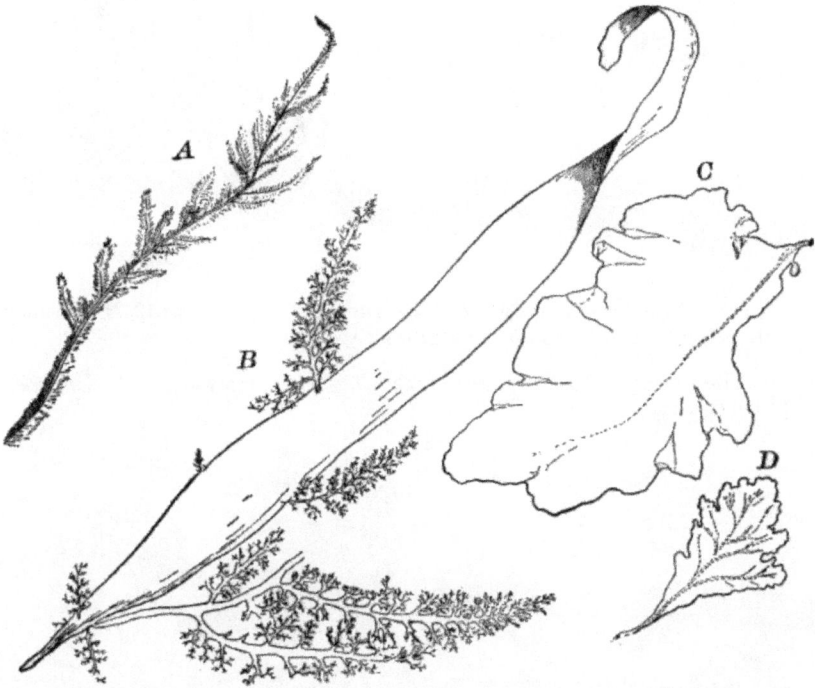

FIG. 30.—*Marine red seaweeds. A, Dasya. B, Rhodymenia (with smaller algæ attached). C, Grinnellia. D, Delesseria. A, B, natural size; the others reduced one-half.*

The fresh-water forms are not common, but may occasionally be met with in mill streams and other running water, attached to stones and woodwork, but are much inferior in size and beauty to the marine species. The red color is not so pronounced, and they are, as a rule, somewhat dull colored.

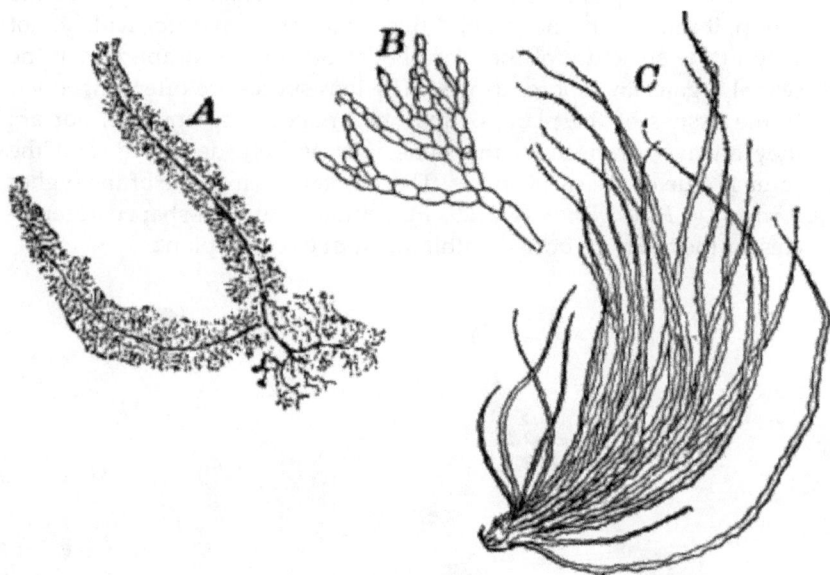

FIG. 31. — *Fresh-water red algæ. A, Batrachospermum, × about 12. B, a branch of the same, × 150. C, Lemanea, natural size.*

The commonest genera are *Batrachospermum* and *Lemanea* (Fig. 31).

CHAPTER VIII.
SUB-KINGDOM III.
Fungi.

THE name "Fungi" has been given to a vast assemblage of plants, varying much among themselves, but on the whole of about the same structural rank as the algæ. Unlike the algæ, however, they are entirely destitute of chlorophyll, and in consequence are dependent upon organic matter for food, some being parasites (growing upon living organisms), others saprophytes (feeding on dead matter). Some of them show close resemblances in structure to certain algæ, and there is reason to believe that they are descended from forms that originally had chlorophyll; others are very different from any green plants, though more or less evidently related among themselves. Recognizing then these distinctions, we may make two divisions of the sub-kingdom: I. The Alga-Fungi (*Phycomycetes*), and II. The True Fungi (*Mycomycetes*).

Class I. — Phycomycetes.

These are fungi consisting of long, undivided, often branching tubular filaments, resembling quite closely those of *Vaucheria* or other *Siphoneæ*, but always destitute of any trace of chlorophyll. The simplest of these include the common moulds (*Mucorini*), one of which will serve to illustrate the characteristics of the order.

If a bit of fresh bread, slightly moistened, is kept under a bell jar or tumbler in a warm room, in the course of twenty-four hours or so it will be covered with a film of fine white threads, and a little later will produce a crop of little globular bodies mounted on upright stalks. These are at first white, but soon become black, and the filaments bearing them also grow dark-colored.

These are moulds, and have grown from spores that are in the atmosphere falling on the bread, which offers the proper conditions for their growth and multiplication.

One of the commonest moulds is the one here figured (Fig. 32), and named *Mucor stolonifer*, from the runners, or "stolons," by which it spreads from one point to another. As it grows it sends out these runners along the surface of the bread, or even along the inner surface of the glass covering it. They fasten themselves at intervals to the substratum, and send up from these points clusters of short filaments, each one tipped with a spore case, or "sporangium."

For microscopical study they are best mounted in dilute glycerine (about one-quarter glycerine to three-quarters pure water). After carefully spreading out the specimens in this mixture, allow a drop of alcohol to fall upon the preparation, and then put on the cover glass. The alcohol drives out the air, which otherwise interferes badly with the examination.

The whole plant consists of a very long, much-branched, but undivided tubular filament. Where it is in contact with the substratum, root-like outgrowths are formed, not unlike those observed in *Vaucheria*. At first the walls are colorless, but later become dark smoky brown in color. A layer of colorless granular protoplasm lines the wall, becoming more abundant toward the growing tips of the branches. The spore cases, "sporangia," arise at the ends of upright branches (Fig. 32, *C*), which at first are cylindrical (*a*), but later enlarge at the end (*b*), and become cut off by a convex wall (*c*). This wall pushes up into the young sporangium, forming a structure called the "columella." When fully grown, the sporangium is globular, and appears quite opaque, owing to the numerous granules in the protoplasm filling the space between the columella and its outer wall. This protoplasm now divides into a great number of small oval cells (spores), which rapidly darken, owing to a thick, black wall formed about each one, and at the same time the columella and the stalk of the sporangium become dark-colored.

When ripe, the wall of the sporangium dissolves, and the spores (Fig. 32, *E*) are set free. The columella remains unchanged, and some of the spores often remain sticking to it (Fig. 32, *D*).

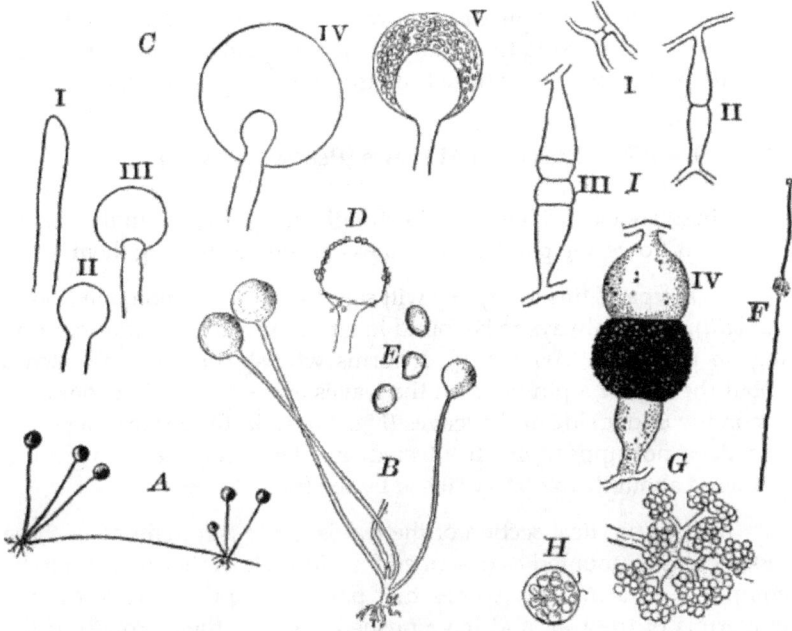

FIG. 32.—A, *common black mould (Mucor),* × 5. B, *three nearly ripe spore cases,* × 25. C, *development of the spore cases,* I–IV, × 150; V, × 50. D, *spore case which has discharged its spores. E, spores,* × 300. F, *a form of Mucor mucedo, with small accessory spore cases,* × 5. G, *the spore cases,* × 50. H, *a single spore case,* × 300. I, *development of the zygospore of a black mould,* × 45 *(after De Bary).*

Spores formed in a manner strongly recalling those of the pond scums are also known, but only occur after the plants have grown for a long time, and hence are rarely met with (Fig. 32, *I*).

Another common mould (*M. mucedo*), often growing in company with the one described, differs from it mainly in the longer stalk of the sporangium, which is also smaller, and in not forming runners. This species sometimes bears clusters of very small sporangia attached to the middle of the ordinary sporangial filament (Fig. 32, *F, H*). These small sporangia have no columella.

Other moulds are sometimes met with, parasitic upon the larger species of *Mucor*.

Related to the black moulds are the insect moulds (*Entomopthoreæ*), which attack and destroy insects. The commonest of these attacks the house flies in autumn, when the flies, thus infested, may often be found sticking to window panes, and surrounded by a whitish halo of the spores that have been thrown off by the fungus.

Order II. — White Rusts and Mildews (*PERONOSPOREÆ*)

These are exclusively parasitic fungi, and grow within the tissues of various flowering plants, sometimes entirely destroying them.

As a type of this group we will select a very common one (*Cystopus bliti*), that is always to be found in late summer and autumn growing on pig weed (*Amarantus*). It forms whitish, blister-like blotches about the size of a pin head on the leaves and stems, being commonest on the under side of the leaves (Fig. 33, *A*). In the earlier stages the leaf does not appear much affected, but later becomes brown and withered about the blotches caused by the fungus.

If a thin vertical section of the leaf is made through one of these blotches, and mounted as described for *Mucor*, the latter is found to be composed of a mass of spores that have been produced below the epidermis of the leaf, and have pushed it up by their growth. If the section is a very thin one, we may be able to make out the structure of the fungus, and then find it to be composed of irregular, tubular, much-branched filaments, which, however, are not divided by cross-walls. These filaments run through the intercellular spaces of the leaf, and send into the cells little globular suckers, by means of which the fungus feeds.

The spores already mentioned are formed at the ends of crowded filaments, that push up, and finally rupture the epidermis (Fig. 33, *B*). They are formed by the ends of the filaments swelling up and becoming constricted, so as to form an oval spore, which is then cut off by a wall. The portion of the filament immediately below acts in the same way, and the process is repeated until a chain of half a dozen or more may be produced, the lowest one being always the last formed. When ripe, the spores are separated by a thin neck, and become very easily broken off.

In order to follow their germination it is only necessary to place a few leaves with fresh patches of the fungus under a bell jar or tumbler, inverted over a dish full of water, so as to keep the air within saturated with moisture, but taking care to keep the leaves out of the water. After about twenty-four hours, if some of the spores are scraped off and mounted in water, they will germinate in the course of an hour or so. The contents divide into about eight parts, which escape from the top of the spore, which at this time projects as a little papilla. On escaping, each mass of protoplasm swims away as a zoöspore, with two extremely delicate cilia. After a short time it comes to rest, and, after developing a thin cell wall, germinates by sending out one or two filaments (Fig. 33, C, E).

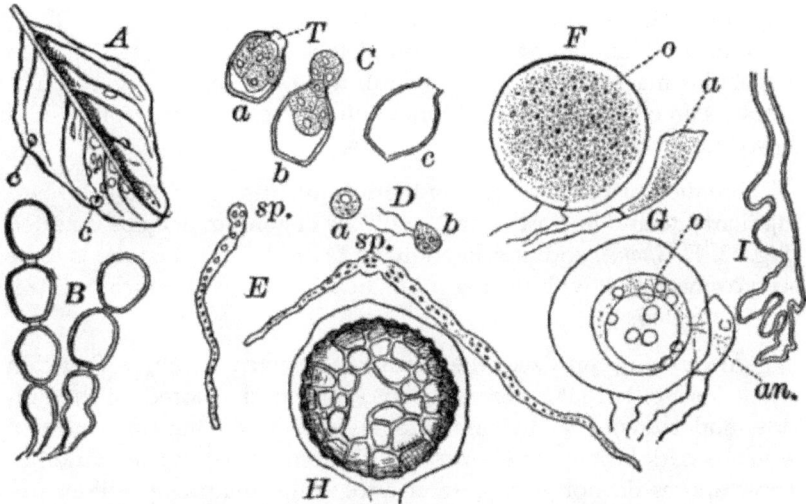

FIG. 33.—A, leaf of pig-weed (Amarantus), with spots of white rust (c), one-half natural size. B, non-sexual spores (conidia). C, the same germinating. D, zoöspores. E, germinating zoöspores. sp. the spore. F, young. G, mature sexual organs. In G, the tube may be seen connecting the antheridium (an.), with the egg cell (o). H, a ripe resting spore still surrounded by the wall of the oögonium. I, a part of a filament of the fungus, showing its irregular form. All × 300.

Under normal conditions the spores probably germinate when the leaves are wet, and the filaments enter the plant through the breathing pores on the lower surface of the leaves, and spread rapidly through the intercellular spaces.

Later on, spores of a very different kind are produced. Unlike those already studied, they are formed some distance below the epidermis, and in order to study them satisfactorily, the fungus must be freed from the host plant. In order to do this, small pieces of the leaf should be boiled for about a minute in strong caustic potash, and then treated with acetic or hydrochloric acid. By this means the tissues of the leaf become so soft as to be readily removed, while the fungus is but little affected. The preparation should now be washed and mounted in dilute glycerine.

The spores (oöspores) are much larger than those first formed, and possess an outer coat of a dark brown color (Fig. 33, *H*). Each spore is contained in a large cell, which arises as a swelling of one of the filaments, and becomes shut off by a wall. At first (Fig. 33, *F*) its contents are granular, and fill it completely, but later contract to form a globular mass of protoplasm (G. *o*), the germ cell or egg cell. The whole is an oögonium, and differs in no essential respect from that of *Vaucheria*.

Frequently a smaller cell (antheridium), arising from a neighboring filament, and in close contact with the oögonium, may be detected (Fig. 33, *F*, *G*, *an.*), and in exceptionally favorable cases a tube is to be seen connecting it with the germ cell, and by means of which fertilization is effected.

After being fertilized, the germ cell secretes a wall, at first thin and colorless, but later becoming thick and dark-colored on the outside, and showing a division into several layers, the outermost of which is dark brown, and covered with irregular reticulate markings. These spores do not germinate at once, but remain over winter unchanged.

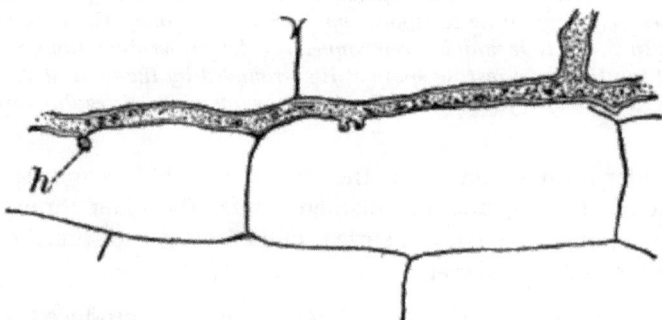

FIG. 34. — *Fragment of a filament of the white rust of the shepherd's-purse, showing the suckers (h),* × 300.

It is by no means impossible that sometimes the germ cell may develop into a spore without being fertilized, as is the case in many of the water moulds.

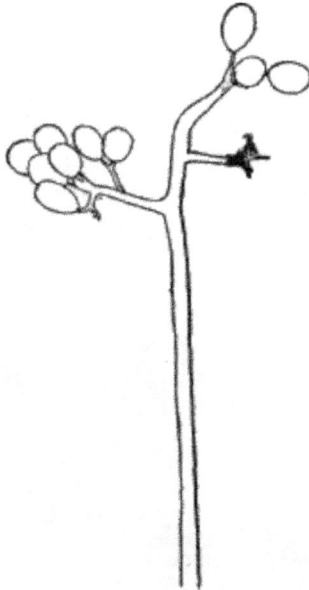

FIG. 35. — *Non-sexual spores of the vine mildew (Peronospora viticola), × 150.*

Closely related to the species above described is another one (*C. candidus*), which attacks shepherd's-purse, radish, and others of the mustard family, upon which it forms chalky white blotches, and distorts the diseased parts of the plant very greatly.

For some reasons this is the best species for study, longitudinal sections through the stem showing very beautifully the structure of the fungus, and the penetration of the cells of the host [4] by the suckers (Fig. 34).

Very similar to the white rusts in most respects, but differing in the arrangement of the non-sexual spores, are the mildews (*Perono-spora, Phytophthora*). These plants form mouldy-looking patches on the leaves and stems of many plants, and are often very destructive. Among them are the vine mildew (*Peronospora viticola*) (Fig. 35), the potato fungus (*Phytophthora infestans*), and many others.

Order III. — *SAPROLEGNIACEÆ* (Water Moulds).

These plants resemble quite closely the white rusts, and are probably related to them. They grow on decaying organic matter in water, or sometimes on living water animals, fish, crustaceans, etc. They may usually be had for study by throwing into water taken from a stagnant pond or aquarium, a dead fly or some other insect. After a few days it will probably be found covered with a dense growth of fine, white filaments, standing out from it in all directions (Fig. 36, *A*). Somewhat later, if carefully examined with a lens, little round, white bodies may be seen scattered among the filaments.

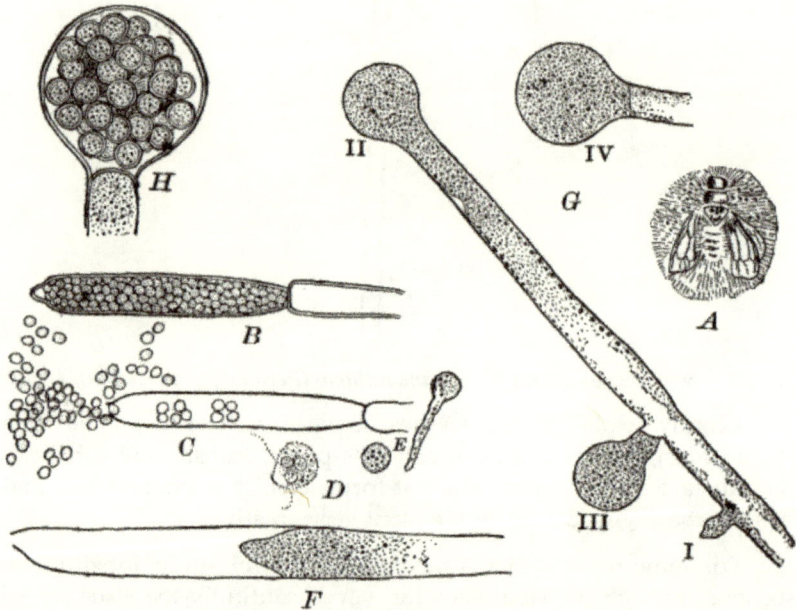

FIG. 36. — *A, an insect that has decayed in water, and become attacked by a water mould (Saprolegnia), natural size. B, a ripe zoösporangium, × 100. C, the same discharging the spores. D, active. E, germinating zoöspores, × 300. F, a second sporangium forming below the empty one. G I–IV, development of the oögonium, × 100. H, ripe oögonium filled with resting spores, × 100.*

On carefully removing a bit of the younger growth and examining it microscopically, it is found to consist of long filaments much like those of *Vaucheria*, but entirely destitute of chlorophyll. In places these filaments are filled with densely granular protoplasm, which when highly magnified exhibits streaming movements. The protoplasm contains a large amount of oil in the form of small, shining drops.

In the early stages of its growth the plant multiplies by zoöspores, produced in great numbers in sporangia at the ends of the branches. The protoplasm collects here much as we saw in *V. sessilis*, the end of the filament becoming club-shaped and ending in a short protuberance (Fig. 36, *B*). This end becomes separated by a wall, and the contents divide into numerous small cells that sometimes are naked, and sometimes have a delicate membrane about them. The first sign of division is the appearance in the protoplasm of delicate lines dividing it into numerous polygonal areas which soon become more distinct, and are seen to be distinct cells whose outlines remain more or less angular on account of the mutual pressure. When ripe, the end of the sporangium opens, and the contained cells are discharged (Fig. 36, *C*). In case they have no membrane, they swim away at once, each being provided with two cilia, and resembling almost exactly the zoöspores of the white rust (Fig. 36, *D*, *E*). When the cells are surrounded by a membrane they remain for some time at rest, but finally the contents escape as a zoöspore, like those already described. By killing the zoöspores with a little iodine the granular nature of the protoplasm is made more evident, and the cilia may be seen. They soon come to rest, and germinate in the same way as those of the white rusts and mildews.

As soon as the sporangium is emptied, a new one is formed, either by the filament growing up through it (Fig. 36, *F*) and the end being again cut off, or else by a branch budding out just below the base of the empty sporangium, and growing up by the side of it.

Besides zoöspores there are also resting spores developed. Oögonia like those of *Vaucheria* or the *Peronosporeæ* are formed usually after the formation of zoöspores has ceased; but in many cases, perhaps all, these develop without being fertilized. Antheridia are often wanting, and even when they are present, it is very doubtful whether fertilization takes place. [5]

The oögonia (Fig. 36, *G*, *H*) arise at the end of the main filaments, or of short side branches, very much as do the sporangia, from which they differ at this stage in being of globular form. The contents contract to form one or several egg cells, naked at first, but later becoming thick-walled resting spores (*H*).

CHAPTER IX.
THE TRUE FUNGI (Mycomycetes).

THE great majority of the plants ordinarily known as *fungi* are embraced under this head. While some of the lower forms show affinities with the *Phycomycetes*, and through them with the algæ, the greater number differ very strongly from all green plants both in their habits and in their structure and reproduction. It is a much-disputed point whether sexual reproduction occurs in any of them, and it is highly probable that in the great majority, at any rate, the reproduction is purely non-sexual.

Probably to be reckoned with the *Mycomycetes*, but of doubtful affinities, are the small unicellular fungi that are the main causes of alcoholic fermentation; these are the yeast fungi (*Saccharomycetes*). They cause the fermentation of beer and wine, as well as the incipient fermentation in bread, causing it to "rise" by the giving off of bubbles of carbonic acid gas during the process.

If a little common yeast is put into water containing starch or sugar, and kept in a warm place, in a short time bubbles of gas will make their appearance, and after a little longer time alcohol may be detected by proper tests; in short, alcoholic fermentation is taking place in the solution.

If a little of the fermenting liquid is examined microscopically, it will be found to contain great numbers of very small, oval cells, with thin cell walls and colorless contents. A careful examination with a strong lens (magnifying from 500–1000 diameters) shows that the protoplasm, in which are granules of varying size, does not fill the cell completely, but that there are one or more large vacuoles or spaces filled with colorless cell sap. No nucleus is visible in the living cell, but it has been shown that a nucleus is present.

If growth is active, many of the cells will be seen dividing. The process is somewhat different from ordinary fission and is called budding (Fig. 37, *B*). A small protuberance appears at the bud or at the side of the cell, and enlarges rapidly, assuming the form of the mother cell, from which it becomes completely separated by the constriction of the base, and may fall off at once, or, as is more frequently the case, may remain attached for a time, giving rise itself to other buds, so that not infrequently groups of half a dozen or more cells are met with (Fig. 37, *B*, *C*).

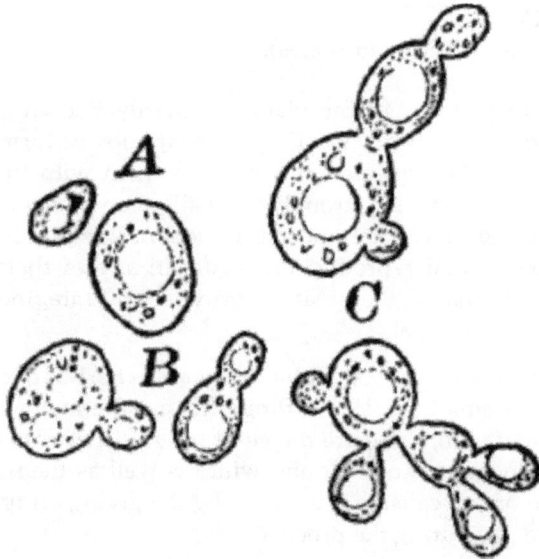

FIG. 37.—*A, single cells of yeast. B, C, similar cells, showing the process of budding, × 750.*

That the yeast cells are the principal agents of alcoholic fermentation may be shown in much the same way that bacteria are shown to cause ordinary decomposition. Liquids from which they are excluded will remain unfermented for an indefinite time.

There has been much controversy as to the systematic position of the yeast fungi, which has not yet been satisfactorily settled, the question being whether they are to be regarded as independent plants or only one stage in the life history of some higher fungi (possibly the *Smuts*), which through cultivation have lost the power of developing further.

Class I.—The Smuts (*USTILLAGINEÆ*).

The smuts are common and often very destructive parasitic fungi, living entirely within the tissues of the higher plants. Owing to this, as well as to the excessively small spores and difficulty in germinating them, the plants are very difficult of study, except in a general way, and we will content ourselves with a glance at one of the common forms, the corn smut (*Ustillago maydis*). This familiar fungus attacks Indian corn, forming its spores in enormous quantities in various parts of the diseased plant, but particularly in the flowers ("tassel" and young ear).

The filaments, which resemble somewhat those of the white rusts, penetrate all parts of the plant, and as the time approaches for the formation of the spores, these branch extensively, and at the same time become soft and mucilaginous (Fig. 38, B). The ends of these short branches enlarge rapidly and become shut off by partitions, and in each a globular spore (Fig. 38, C) is produced. The outer wall is very dark-colored and provided with short spines. To study the filaments and spore formation, very thin sections should be made through the young kernels or other parts in the vicinity, before they are noticeably distorted by the growth of the spore-bearing filaments.

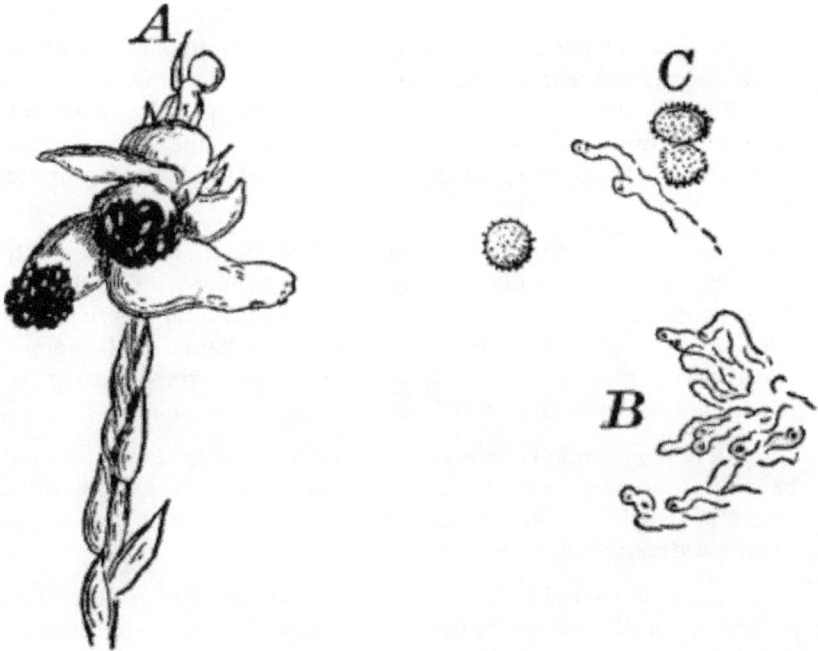

FIG. 38.—A, "tassel" of corn attacked by smut (Ustillago). B, filaments of the fungus from a thin section of a diseased grain, showing the beginning of the formation of the spores, × 300. C, ripe spores, × 300.

As the spores are forming, an abnormal growth is set up in the cells of the part attacked, which in consequence becomes enormously enlarged (Fig. 38, *A*), single grains sometimes growing as large as a walnut. As the spores ripen, the affected parts, which are at first white, become a livid gray, due to the black spores shining through the overlying white tissues. Finally the masses of spores burst through the overlying cells, appearing like masses of soot, whence the popular name for the plant.

The remaining *Mycomycetes* are pretty readily divisible into two great classes, based upon the arrangement of the spores. The first of these is known as the *Ascomycetes* (Sac fungi), the other the *Basidiomycetes* (mushrooms, puff-balls, etc.).

Class II.—*ASCOMYCETES* (Sac Fungi).

This class includes a very great number of common plants, all resembling each other in producing spores in sacs (*asci*, sing. *ascus*) that are usually oblong in shape, and each containing eight spores, although the number is not always the same. Besides the spores formed in these sacs (ascospores), there are other forms produced in various ways.

There are two main divisions of the class, the first including only a few forms, most of which are not likely to be met with by the student. In these the spore sacs are borne directly upon the filaments without any protective covering. The only form that is at all common is a parasitic fungus (*Exoascus*) that attacks peach-trees, causing the disease of the leaves known as "curl."

All of the common *Ascomycetes* belong to the second division, and have the spore sacs contained in special structures called spore fruits, that may reach a diameter of several centimetres in a few cases, though ordinarily much smaller.

Among the simpler members of this group are the mildews (*Perisporiaceæ*), mostly parasitic forms, living upon the leaves and stems of flowering plants, sometimes causing serious injury by their depredations. They form white or grayish downy films on the surface of the plant, in certain stages looking like hoar-frost. Being very common, they may be readily obtained, and are easily studied. One of the best species for study (*Podosphæra*) grows abundantly on the leaves of the dandelion, especially when the plants are growing under unfavorable conditions. The same species is also found on other plants of the same family. It may be found at almost any time during the summer; but for studying, the spore fruits material should be collected in late summer or early autumn. It at first appears as white, frost-like patches, growing dingier as it becomes older, and careful scrutiny of the older specimens will show numerous brown or blackish specks scattered over the patches. These are the spore fruits.

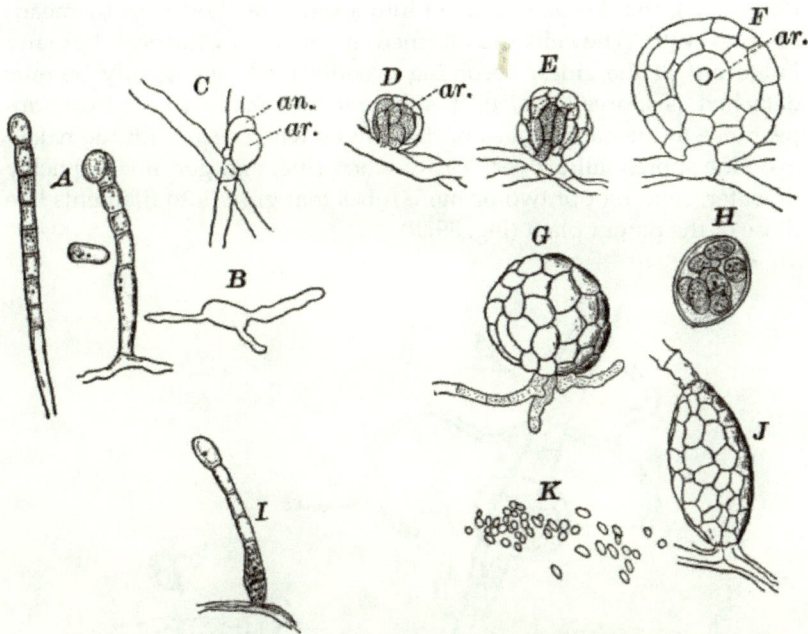

FIG. 39.—A, spore-bearing filaments of the dandelion mildew (Podosphæra), × 150. B, a germinating spore, × 150. C–F, development of the spore fruit, × 300. ar. archicarp. G, a ripe spore fruit, × 150. H, the spore sac removed from the spore fruit, × 150. I, spore-bearing filament attacked by another fungus (Cicinnobulus), causing the enlargement of the basal cell, × 150. J, a more advanced stage, × 300. K, spores, × 300.

For microscopical study, fresh material may be used, or, if necessary, dried specimens. The latter, before mounting, should be soaked for a short time in water, to which has been added a few drops of caustic-potash solution. This will remove the brittleness, and swell up the dried filaments to their original proportions. A portion of the plant should be carefully scraped off the leaf on which it is growing, thoroughly washed in pure water, and transferred to a drop of water or very dilute glycerine, in which it should be carefully spread out with needles. If air bubbles interfere with the examination, they may be driven off with alcohol, and then the cover glass put on. If the specimen is mounted in glycerine, it will keep indefinitely, if care is taken to seal it up. The plant consists of much-interlaced filaments, divided at intervals by cross-walls. [6] They are nearly colorless, and the contents are not conspicuous. These filaments send up vertical branches (Fig. 39, A), that become divided into a series of short cells by means of cross-walls. The cells thus formed are at first cylindrical, but later bulge out at the sides, becoming broadly oval, and finally become detached as spores (*conidia*). It is these spores that give the frosty appearance to the early stages of the fungus when seen with the naked eye. The spores fall off very easily when ripe, and germinate quickly in water, sending out two or more tubes that grow into filaments like those of the parent plant (Fig. 39, B).

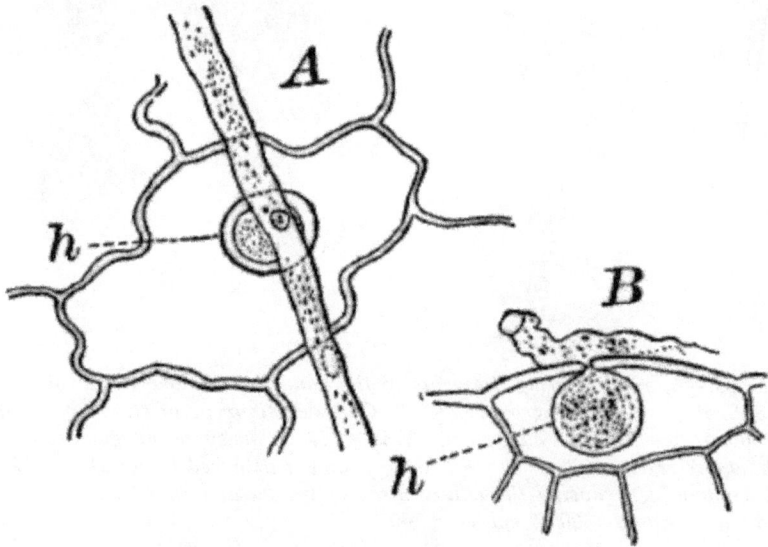

FIG. 40. — *Chrysanthemum mildew (Erysiphe), showing the suckers (h) by which the filaments are attached to the leaf. A, surface view. B, vertical section of the leaf,* × 300.

The spore fruits, as already observed, are formed toward the end of the season, and, in the species under consideration at least, appear to be the result of a sexual process. The sexual organs (if they are really such) are extremely simple, and, owing to their very small size, are not easily found. They arise as short branches at a point where two filaments cross; one of them (Fig. 39, C, ar.), the female cell, or "archi-carp," is somewhat larger than the other and nearly oval in form, and soon becomes separated by a partition from the filament that bears it. The other branch (antheridium) grows up in close contact with the archicarp, and like it is shut off by a partition from its filament. It is more slender than the archicarp, but otherwise differs little from it. No actual communication can be shown to be present between the two cells, and it is therefore still doubtful whether fertilization really takes place. Shortly after these organs are full-grown, several short branches grow up about them, and soon completely envelop them (D, E). These branches soon grow together, and cross-walls are formed in them, so that the young spore fruit appears surrounded by a single layer of cells, sufficiently transparent, however, to allow a view of the interior.

The antheridium undergoes no further change, but the archicarp soon divides into two cells, — a small basal one and a larger upper cell. There next grow from the inner surface of the covering cells, short filaments, that almost completely fill the space between the archicarp and the wall. An optical section of such a stage (Fig. 39, F) shows a double wall and the two cells of the archicarp. The spore fruit now enlarges rapidly, and the outer cells become first yellow and then dark brown, the walls becoming thicker and harder as they change color. Sometimes special filaments or appendages grow out from their outer surfaces, and these are also dark-colored. Shortly before the fruit is ripe, the upper cell of the archicarp, which has increased many times in size, shows a division of its contents into eight parts, each of which develops a wall and becomes an oval spore. By crushing the ripe spore fruit, these spores still enclosed in the mother cell (ascus) may be forced out (Fig. 39, H). These spores do not germinate at once, but remain dormant until the next year.

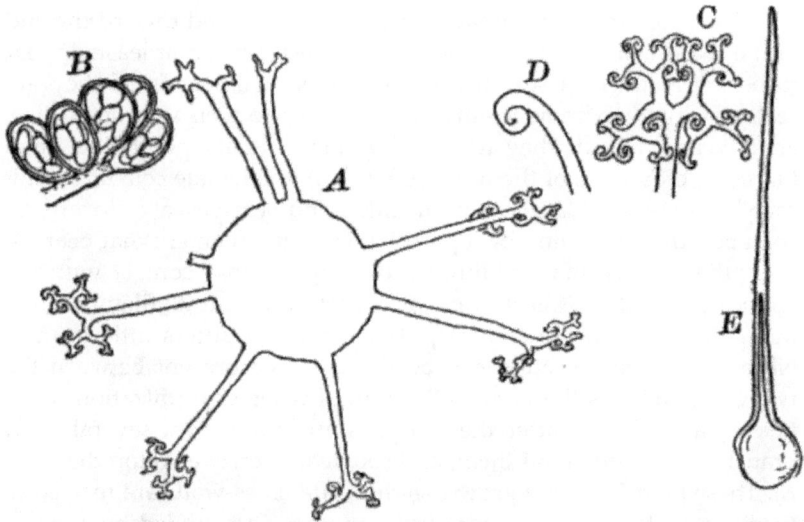

FIG. 41. — *Forms of mildews (Erysiphe). A, Microsphæra, a spore fruit,* × *150. B, cluster of spore sacs of the same,* × *150. C, a single appendage,* × *300. D, end of an appendage of Uncinula,* × *300. E, appendage of Phyllactinia,* × *150.*

Frequently other structures, resembling somewhat the spore fruits, are found associated with them (Fig. 39, *I, K*), and were for a long time supposed to be a special form of reproductive organ; but they are now known to belong to another fungus (*Cicinnobulus*), parasitic upon the mildew. They usually appear at the base of the chains of conidia, causing the basal cell to enlarge to many times its original size, and finally kill the young conidia, which shrivel up. A careful examination reveals the presence of very fine filaments within those of the mildew, which may be traced up to the base of the conidial branch, where the receptacle of the parasite is forming. The spores contained in these receptacles are very small (Fig. 39, *K*), and when ripe exude in long, worm-shaped masses, if the receptacle is placed in water.

The mildews may be divided into two genera: *Podosphæra*, with a single ascus in the spore fruit; and *Erysiphe*, with two or more. In the latter the archicarp branches, each branch bearing a spore sac (Fig. 41, *B*).

The appendages growing out from the wall of the spore fruit are often very beautiful in form, and the two genera given above are often subdivided according to the form of these appendages.

A common mould closely allied to the mildews is found on various articles of food when allowed to remain damp, and is also very common on botanical specimens that have been poorly dried, and hence is often called "herbarium mould" (*Eurotium herbariorum*).

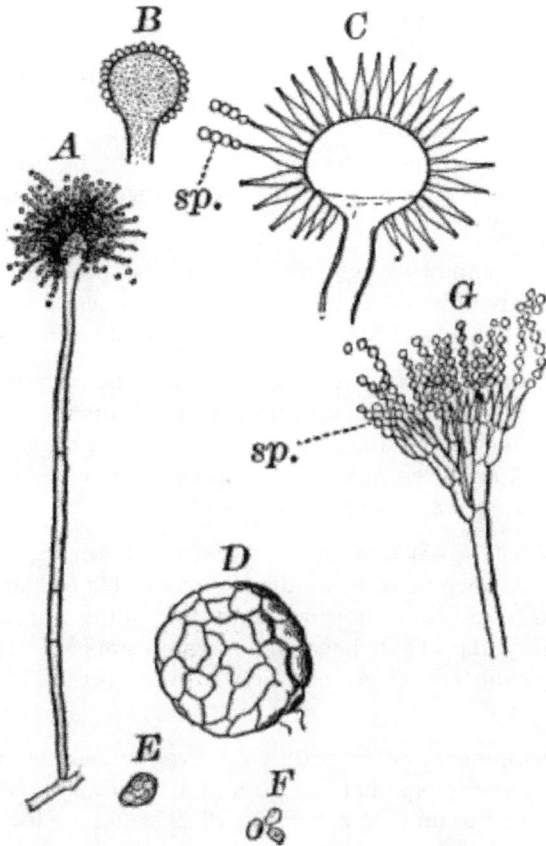

FIG. 42.—A, spore bearing filament of the herbarium mould (Eurotium), × 150. B, C, another species showing the way in which the spores are borne—optical section—× 150. D, spore fruit of the herbarium mould, × 150. E, spore sac. F, spores, × 300. G, spore-bearing filament of the common blue mould (Penicillium), × 300. sp. the spores.

The conidia are of a greenish color, and produced on the ends of upright branches which are enlarged at the end, and from which grow out little prominences, which give rise to the conidia in the same way as we have seen in the mildews (Fig. 42, A).

Spore fruits much like those of the mildews are formed later, and are visible to the naked eye as little yellow grains (Fig. 42, D). These contain numerous very small spore sacs (E), each with eight spores.

There are numerous common species of *Eurotium*, differing in color and size, some being yellow or black, and larger than the ordinary green form.

Another form, common everywhere on mouldy food of all kinds, as well as in other situations, is the blue mould (*Penicillium*). This, in general appearance, resembles almost exactly the herbarium mould, but is immediately distinguishable by a microscopic examination (Fig. 42, G).

In studying all of these forms, they may be mounted, as directed for the black moulds, in dilute glycerine; but must be handled with great care, as the spores become shaken off with the slightest jar.

Of the larger *Ascomycetes*, the cup fungi (*Discomycetes*) may be taken as types. The spore fruit in these forms is often of considerable size, and, as their name indicates, is open, having the form of a flat disc or cup. A brief description of a common one will suffice to give an idea of their structure and development.

Ascobolus (Fig. 43) is a small, disc-shaped fungus, growing on horse dung. By keeping some of this covered with a bell jar for a week or two, so as to retain the moisture, at the end of this time a large crop of the fungus will probably have made its appearance. The part visible is the spore fruit (Fig. 43, A), of a light brownish color, and about as big as a pin-head.

Its development may be readily followed by teasing out in water the youngest specimens that can be found, taking care to take up a little of the substratum with it, as the earliest stages are too small to be visible to the naked eye. The spore fruits arise from filaments not unlike those of the mildews, and are preceded by the formation of an archicarp composed of several cells, and readily seen through the walls of the young fruit (Fig. 43, B). In the study of the early stages, a potash solution will be found useful in rendering them transparent.

The young fruit has much the same structure as that of the mildews, but the spore sacs are much more numerous, and there are special sterile filaments developed between them. If the young spore fruit is treated with chlor-iodide of zinc, it is rendered quite transparent, and the young spore sacs colored a beautiful blue, so that they are readily distinguishable.

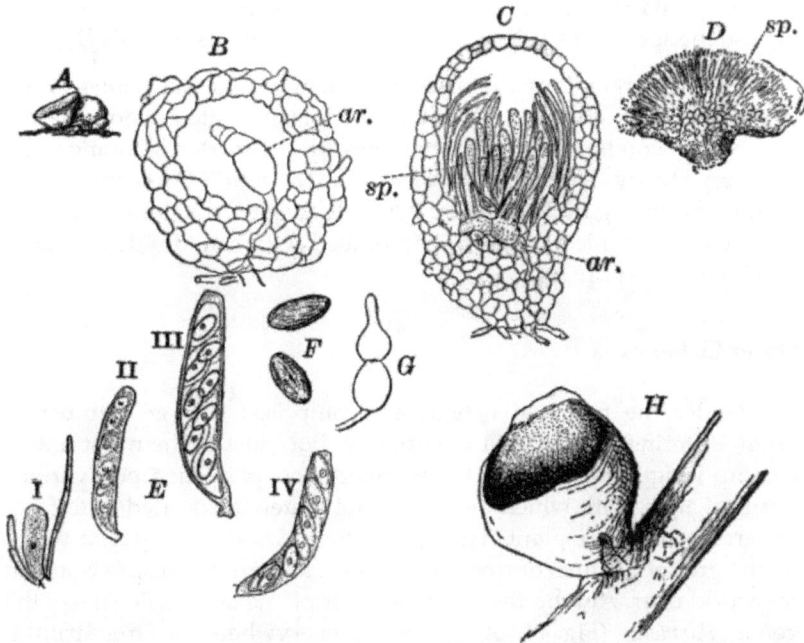

FIG. 43. — *A, a small cup fungus (Ascobolus), × 5. B, young spore fruit, × 300. ar. archicarp. C, an older one, × 150. ar. archicarp. sp. young spore sacs. D, section through a full-grown spore fruit (partly diagrammatic), × 25. sp. spore sacs. E, development of spore sacs and spores: I–III, × 300; IV, × 150. F, ripe spores. G, a sterile filament (paraphysis), × 300. H, large scarlet cup fungus (Peziza), natural size.*

The development of the spore sacs may be traced by carefully crushing the young spore fruits in water. The young spore sacs (Fig. 43, E I) are colorless, with granular protoplasm, in which a nucleus can often be easily seen. The nucleus subsequently divides repeatedly, until there are eight nuclei, about which the protoplasm collects to form as many oval masses, each of which develops a wall and becomes a spore (Figs. II–IV). These are imbedded in protoplasm, which is at first granular, but afterwards becomes almost transparent. As the spores ripen, the wall acquires a beautiful violet-purple color, changing later to a dark purple-brown, and marked with irregular longitudinal ridges (Fig. 43, *F*). The full-grown spore sacs (Fig. 43, *E*, *W*) are oblong in shape, and attached by a short stalk. The sterile filaments between them often become curiously enlarged at the end (*G*). As the spore fruit ripens, it opens at the top, and spreads out so as to expose the spore sacs as they discharge their contents (Fig. 43, *D*).

Of the larger cup fungi, those belonging to the genus *Peziza* (Fig. 43, *H*) are common, growing on bits of rotten wood on the ground in woods. They are sometimes bright scarlet or orange-red, and very showy. Another curious form is the morel (*Morchella*), common in the spring in dry woods. It is stalked like a mushroom, but the surface of the conical cap is honeycombed with shallow depressions, lined with the spore sacs.

Order Lichenes.

Under the name of lichens are comprised a large number of fungi, differing a good deal in structure, but most of them not unlike the cup fungi. They are, with few exceptions, parasitic upon various forms of algæ, with which they are so intimately associated as to form apparently a single plant. They grow everywhere on exposed rocks, on the ground, trunks of trees, fences, etc., and are found pretty much the world over. Among the commonest of plants are the lichens of the genus *Parmelia* (Fig. 44, *A*), growing everywhere on tree trunks, wooden fences, etc., forming gray, flattened expansions, with much indented and curled margins. When dry, the plant is quite brittle, but on moistening becomes flexible, and at the same time more or less decidedly green in color. The lower surface is white or brown, and often develops root-like processes by which it is fastened to the substratum. Sometimes small fragments of the plant become detached in such numbers as to form a grayish powder over certain portions of it. These, when supplied with sufficient moisture, will quickly produce new individuals.

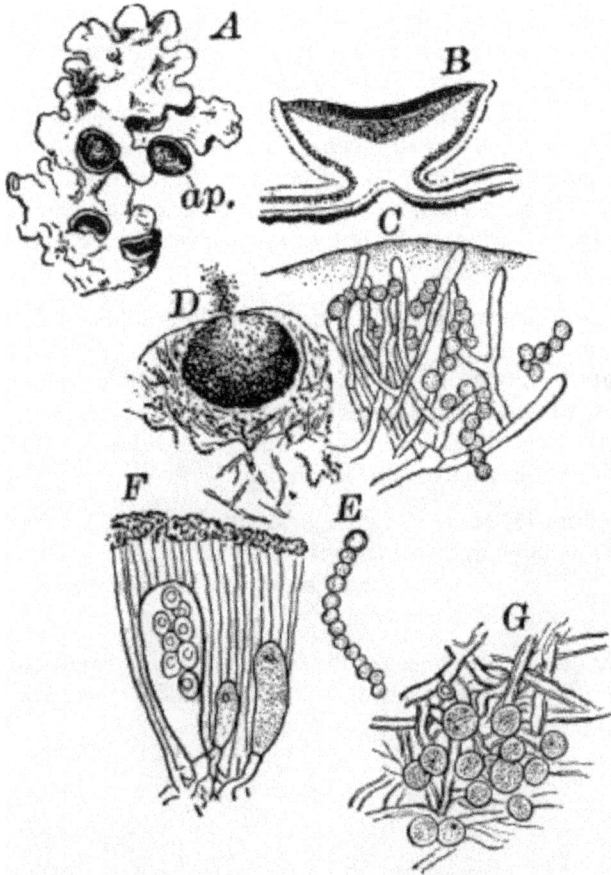

FIG. 44.—A, a common lichen (Parmelia), of the natural size. ap. spore fruit.
B, section through one of the spore fruits, × 5. C, section through the body of a
gelatinous lichen (Collema), showing the Nostoc individuals surrounded by
the fungus filaments, × 300. D, a spermagonium of Collema, × 25. E, a single
Nostoc thread. F, spore sacs and paraphyses of Usnea, × 300. G, Protococcus
cells and fungus filaments of Usnea.

Not infrequently the spore fruits are to be met with flat discs of a
reddish brown color, two or three millimetres in diameter, and closely
resembling a small cup fungus. They are at first almost closed, but
expand as they mature (Fig. 44, A, ap.).

If a thin vertical section of the plant is made and sufficiently
magnified, it is found to be made up of somewhat irregular, thick-

walled, colorless filaments, divided by cross-walls as in the other sac-fungi. In the central parts of the plant these are rather loose, but toward the outside become very closely interwoven and often grown together, so as to form a tough rind. Among the filaments of the outer portion are numerous small green cells, that closer examination shows to be individuals of *Protococcus*, or some similar green algæ, upon which the lichen is parasitic. These are sufficiently abundant to form a green line just inside the rind if the section is examined with a simple lens (Fig. 44, *B*).

The spore fruits of the lichens resemble in all essential respects those of the cup fungi, and the spore sacs (Fig. 44, *F*) are much the same, usually, though not always, containing eight spores, which are sometimes two-celled. The sterile filaments between the spore sacs usually have thickened ends, which are dark-colored, and give the color to the inner surface of the spore fruit.

In Figure 45, *H*, is shown one of the so-called "*Soredia*," [7] a group of the algæ, upon which the lichen is parasitic, surrounded by some of the filaments, the whole separating spontaneously from the plant and giving rise to a new one.

Owing to the toughness of the filaments, the finer structure of the lichens is often difficult to study, and free use of caustic potash is necessary to soften and make them manageable.

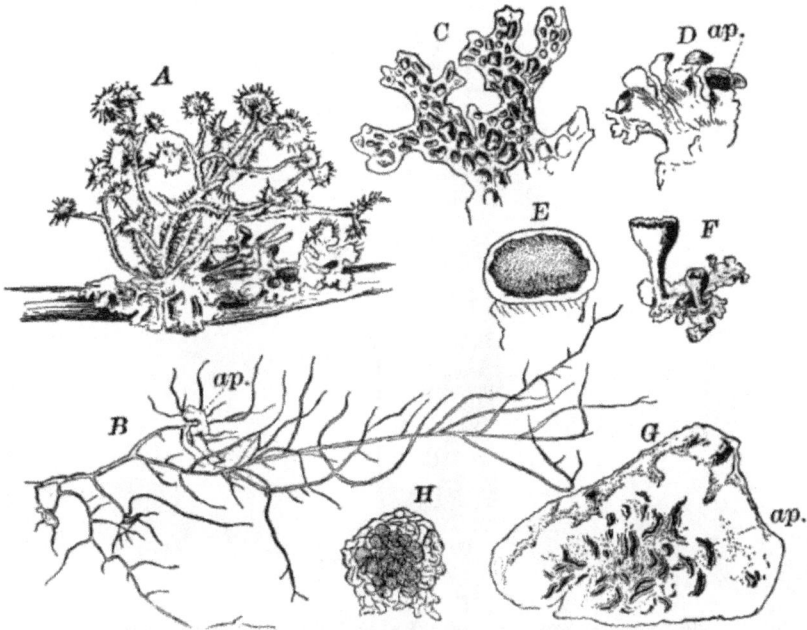

FIG. 45.—*Forms of lichens. A, a branch with lichens growing upon it, one-half natural size. B, Usnea, natural size. ap. spore fruit. C, Sticta, one-half natural size. D, Peltigera, one-half natural size. ap. spore fruit. E, a single spore fruit, × 2. F, Cladonia, natural size. G, a piece of bark from a beech, with a crustaceous lichen (Graphis) growing upon it, × 2. ap. spore fruit. H, Soredium of a lichen, × 300.*

According to their form, lichens are sometimes divided into the bushy (fruticose), leafy (frondose), incrusting (crustaceous), and gelatinous. Of the first, the long gray *Usnea* (Fig. 45, *A, B*), which drapes the branches of trees in swamps, is a familiar example; of the second, *Parmelia, Sticta* (Fig. 45, *C*) and *Peltigera* (*D*) are types; of the third, *Graphis* (*G*), common on the trunks of beech-trees, to which it closely adheres; and of the last, *Collema* (Fig. 44, *C, D, E*), a dark greenish, gelatinous form, growing on mossy tree trunks, and looking like a colony of *Nostoc*, which indeed it is, but differing from an ordinary colony in being penetrated everywhere by the filaments of the fungus growing upon it.

FIG. 46. — Branch of a plum-tree attacked by black knot. Natural size.

Not infrequently in this form, as well as in other lichens, special cavities, known as spermogonia (Fig. 44, *D*), are found, in which excessively small spores are produced, which have been claimed to be male reproductive cells, but the latest investigations do not support this theory.

The last group of the *Ascomycetes* are the "black fungi," *Pyreno-mycetes*, represented by the black knot of cherry and plum trees, shown in Figure 46. They are mainly distinguished from the cup fungi by producing their spore sacs in closed cavities. Some are parasites; others live on dead wood, leaves, etc., forming very hard masses, generally black in color, giving them their common name. Owing to the hardness of the masses, they are very difficult to manipulate; and, as the structure is not essentially different from that of the *Discomycetes*, the details will not be entered into here.

Of the parasitic forms, one of the best known is the "ergot" of rye, more or less used in medicine. Other forms are known that attack insects, particularly caterpillars, which are killed by their attacks.

CHAPTER X.
FUNGI — *Continued.*

Class Basidiomycetes.

THE *Basidiomycetes* include the largest and most highly developed of the fungi, among which are many familiar forms, such as the mushrooms, toadstools, puff-balls, etc. Besides these large and familiar forms, there are other simpler and smaller ones that, according to the latest investigations, are probably related to them, though formerly regarded as constituting a distinct group. The most generally known of these lower *Basidiomycetes* are the so-called rusts. The larger *Basidiomycetes* are for the most part saprophytes, living in decaying vegetable matter, but a few are true parasites upon trees and others of the flowering plants.

All of the group are characterized by the production of spores at the top of special cells known as basidia, [8] the number produced upon a single basidium varying from a single one to several.

Of the lower *Basidiomycetes*, the rusts (*Uredineæ*) offer common and easily procurable forms for study. They are exclusively parasitic in their habits, growing within the tissues of the higher land plants, which they often injure seriously. They receive their popular name from the reddish color of the masses of spores that, when ripe, burst through the epidermis of the host plant. Like many other fungi, the rusts have several kinds of spores, which are often produced on different hosts; thus one kind of wheat rust lives during part of its life within the leaves of the barberry, where it produces spores quite different from those upon the wheat; the cedar rust, in the same way, is found at one time attacking the leaves of the wild crab-apple and thorn.

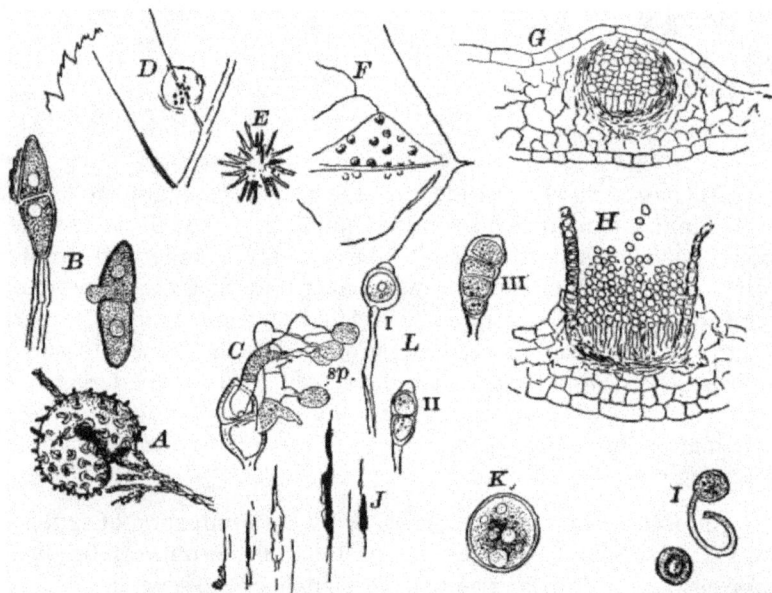

FIG. 47.—A, a branch of red cedar attacked by a rust (Gymnosporangium), causing a so-called "cedar apple," × ½. B, spores of the same, one beginning to germinate, × 300. C, a spore that has germinated, each cell producing a short, divided filament (basidium), which in turn gives rise to secondary spores (sp.), × 300. D, part of the leaf of a hawthorn attacked by the cluster cup stage of the same fungus, upper side showing spermogonia, natural size. E, cluster cups (Roestelia) of the same fungus, natural size. F, tip of a leaf of the Indian turnip (Arisæma), bearing the cluster cup (Æcidium) stage of a rust, × 2. G, vertical section through a young cluster cup. H, similar section through a mature one, × 50. I, germinating spores of H, × 300. J, part of a corn leaf, with black rust, natural size. K, red rust spore of the wheat rust (Puccinia graminis), × 300. L, forms of black-rust spores: I, Uromyces; II, Puccinia; III, Phragmidium.

The first form met with in most rusts is sometimes called the "cluster-cup" stage, and in many species is the only stage known. In Figure 47, F, is shown a bit of the leaf of the Indian turnip (Arisæma) affected by one of these "cluster-cup" forms. To the naked eye, or when slightly magnified, the masses of spores appear as bright orange spots, mostly upon the lower surface. The affected leaves are more or less checked in their growth, and the upper surface shows lighter blotches, corresponding to the areas below that bear the cluster cups. These at first appear as little elevations of a yellowish color, and

91

covered with the epidermis; but as the spores ripen they break through the epidermis, which is turned back around the opening, the whole forming a little cup filled with a bright orange red powder, composed of the loose masses of spores.

Putting a piece of the affected leaf between two pieces of pith so as to hold it firmly, with a little care thin vertical sections of the leaf, including one of the cups, may be made, and mounted, either in water or glycerine, removing the air with alcohol. We find that the leaf is thickened at this point owing to a diseased growth of the cells of the leaf, induced by the action of the fungus. The mass of spores (Fig. 47, G) is surrounded by a closely woven mass of filaments, forming a nearly globular cavity. Occupying the bottom of the cup are closely set, upright filaments, each bearing a row of spores, arranged like those of the white rusts, but so closely crowded as to be flattened at the sides. The outer rows have thickened walls, and are grown together so as to form the wall of the cup.

The spores are filled with granular protoplasm, in which are numerous drops of orange-yellow oil, to which is principally due their color. As the spores grow, they finally break the overlying epidermis, and then become rounded as the pressure from the sides is relieved. They germinate within a few hours if placed in water, sending out a tube, into which pass the contents of the spore (Fig. 47, I).

One of the most noticeable of the rusts is the cedar rust (*Gymnosporangium*), forming the growths known as "cedar apples," often met with on the red cedar. These are rounded masses, sometimes as large as a walnut, growing upon the small twigs of the cedar (Fig. 47, A). This is a morbid growth of the same nature as those produced by the white rusts and smuts. If one of these cedar apples is examined in the late autumn or winter, it will be found to have the surface dotted with little elevations covered by the epidermis, and on removing this we find masses of forming spores. These rupture the epidermis early in the spring, and appear then as little spikes of a rusty red color. If they are kept wet for a few hours, they enlarge rapidly by the absorption of water, and may reach a length of four or five centimetres, becoming gelatinous in consistence, and sometimes almost entirely hiding the surface of the "apple." In this stage the fungus is extremely conspicuous, and may frequently be met with after rainy weather in the spring.

This orange jelly, as shown by the microscope, is made up of elongated two-celled spores (teleuto spores), attached to long gelatinous stalks (Fig. 47, *B*). They are thick-walled, and the contents resemble those of the cluster-cup spores described above.

To study the earlier stages of germination it is best to choose specimens in which the masses of spores have not been moistened. By thoroughly wetting these, and keeping moist, the process of germination may be readily followed. Many usually begin to grow within twenty-four hours or less. Each cell of the spore sends out a tube (Fig. 47, *C*), through an opening in the outer wall, and this tube rapidly elongates, the spore contents passing into it, until a short filament (basidium) is formed, which then divides into several short cells. Each cell develops next a short, pointed process, which swells up at the end, gradually taking up all the contents of the cell, until a large oval spore (*sp.*) is formed at the tip, containing all the protoplasm of the cell.

Experiments have been made showing that these spores do not germinate upon the cedar, but upon the hawthorn or crab-apple, where they produce the cluster-cup stage often met with late in the summer. The affected leaves show bright orange-yellow spots about a centimetre in diameter (Fig. 47, *D*), and considerably thicker than the other parts of the leaf. On the upper side of these spots may be seen little black specks, which microscopic examination shows to be spermogonia, resembling those of the lichens. Later, on the lower surface, appear the cluster cups, whose walls are prolonged so that they form little tubular processes of considerable length (Fig. 47, *E*).

In most rusts the teleuto spores are produced late in the summer or autumn, and remain until the following spring before they germinate. They are very thick-walled, the walls being dark-colored, so that in mass they appear black, and constitute the "black-rust" stage (Fig. 47, *J*). Associated with these, but formed earlier, and germinating immediately, are often to be found large single-celled spores, borne on long stalks. They are usually oval in form, rather thin-walled, but the outer surface sometimes provided with little points. The contents are reddish, so that in mass they appear of the color of iron rust, and cause the "red rust" of wheat and other plants, upon which they are growing.

The classification of the rusts is based mainly upon the size and shape of the teleuto spores where they are known, as the cluster-cup and red-rust stages are pretty much the same in all. Of the commoner genera *Melampsora*, and *Uromyces* (Fig. 47, *L* I), have unicellular teleuto spores; *Puccinia* (II) and *Gymnosporangium*, two-celled spores; *Triphragmium*, three-celled; and *Phragmidium* (III), four or more.

The rusts are so abundant that a little search can scarcely fail to find some or all of the stages. The cluster-cup stages are best examined fresh, or from alcoholic material; the teleuto spores may be dried without affecting them.

Probably the best-known member of the group is the wheat rust (*Puccinia graminis*), which causes so much damage to wheat and sometimes to other grains. The red-rust stage may be found in early summer; the black-rust spores in the stubble and dead leaves in the autumn or spring, forming black lines rupturing the epidermis.

Probably to be associated with the lower *Basidiomycetes* are the large fungi of which *Tremella* (Fig. 51, *A*) is an example. They are jelly-like forms, horny and somewhat brittle when dry, but becoming soft when moistened. They are common, growing on dead twigs, logs, etc., and are usually brown or orange-yellow in color.

Of the higher *Basidiomycetes*, the toadstools, mushrooms, etc., are the highest, and any common form will serve for study. One of the most accessible and easily studied forms is *Coprinus*, of which there are several species growing on the excrement of various herbivorous animals. They not infrequently appear on horse manure that has been kept covered with a glass for some time, as described for *Ascobolus*. After two or three weeks some of these fungi are very likely to make their appearance, and new ones continue to develop for a long time.

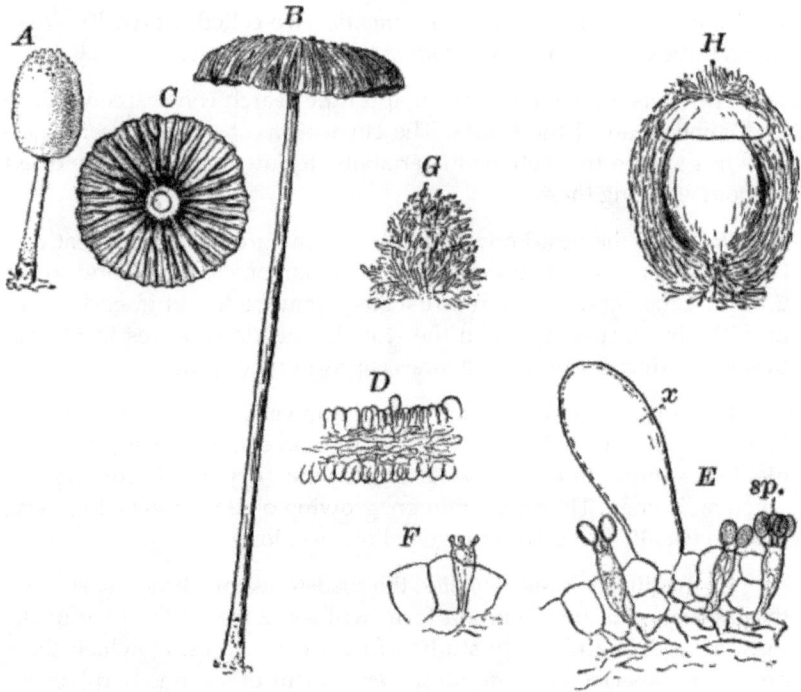

FIG. 48. — *A, young. B, full-grown fruit of a toadstool (Coprinus), × 2. C, under side of the cap, showing the radiating "gills," or spore-bearing plates. D, section across one of the young gills, × 150. E, F, portions of gills from a nearly ripe fruit, × 300. sp. spores. x, sterile cell. In F, a basidium is shown, with the young spores just forming. G, H, young fruits, × 50.*

The first trace of the plant, visible to the naked eye, is a little downy, white speck, just large enough to be seen. This rapidly increases in size, becoming oblong in shape, and growing finally somewhat darker in color; and by the time it reaches a height of a few millimetres a short stalk becomes perceptible, and presently the whole assumes the form of a closed umbrella. The top is covered with little prominences, that diminish in number and size toward the bottom. After the cap reaches its full size, the stalk begins to grow, slowly at first, but finally with great rapidity, reaching a height of several centimetres within a few hours. At the same time that the stalk is elongating, the cap spreads out, radial clefts appearing on its upper surface, which flatten out very much as the folds of an umbrella are stretched as it opens, and the spaces between the clefts appear as ridges, comparable to the ribs of the umbrella (Fig. 48, *B*). The under side of the cap has a number of ridges running from the centre to the margin, and of a black color, due to the innumerable spores covering their surface (*C*). Almost as soon as the umbrella opens, the spores are shed, and the whole structure shrivels up and dissolves, leaving almost no trace behind.

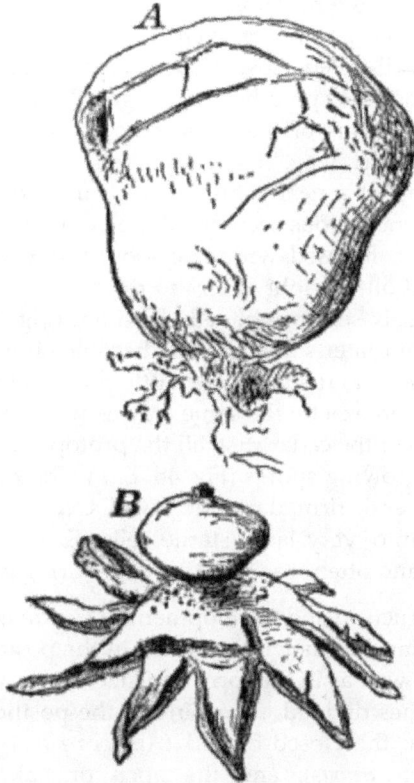

FIG. 49. — *Basidiomycetes. A, common puff-ball (Lycoperdon). B, earth star (Geaster). A, × ¼. B, one-half natural size.*

If we examine microscopically the youngest specimens procurable, freeing from air with alcohol, and mounting in water or dilute glycerine, we find it to be a little, nearly globular mass of colorless filaments, with numerous cross-walls, the whole arising from similar looser filaments imbedded in the substratum (Fig. 48, *G*). If the specimen is not too young, a denser central portion can be made out, and in still older ones (Fig. 48, *H*) this central mass has assumed the form of a short, thick stalk, crowned by a flat cap, the whole invested by a loose mass of filaments that merge more or less gradually into the central portion. By the time the spore fruit (for this structure corresponds to the spore fruit of the *Ascomycetes*) reaches a height of two or three millimetres, and is plainly visible to the naked eye, the cap grows downward at the margins, so as to almost entirely conceal the stalk. A longitudinal section of such a stage shows the stalk to be composed of a small-celled, close tissue becoming looser in the cap, on whose inner surface the spore-bearing ridges ("gills" or *Lamellæ*) have begun to develop. Some of these run completely to the edge of the cap, others only part way. To study their structure, make cross-sections of the cap of a nearly full-grown, but unopened, specimen, and this will give numerous sections of the young gills. We find them to be flat plates, composed within of loosely interwoven filaments, whose ends stand out at right angles to the surface of the gills, forming a layer of closely-set upright cells (basidia) (Fig. 48, *D*). These are at first all alike, but later some of them become club-shaped, and develop at the end several (usually four) little points, at the end of which spores are formed in exactly the same way as we saw in the germinating teleuto spores of the cedar rust, all the protoplasm of the basidium passing into the growing spores (Fig. 48, *E, F*). The ripe spores (*E, sp.*) are oval, and possess a firm, dark outer wall. Occasionally some of the basidia develop into very large sterile cells (*E, x*), projecting far beyond the others, and often reaching the neighboring gill.

Similar in structure and development to *Coprinus* are all the large and common forms; but they differ much in the position of the spore-bearing tissue, as well as in the form and size of the whole spore fruit. They are sometimes divided, according to the position of the spores, into three orders: the closed-fruited (*Angiocarpous*) forms, the half-closed (*Hemi-angiocarpous*), and the open or naked-fruited forms (*Gymnocarpous*).

Of the first, the puff-balls (Fig. 49) are common examples. One species, the giant puff-ball (*Lycoperdon giganteum*), often reaches a diameter of thirty to forty centimetres. The earth stars (*Geaster*) have a double covering to the spore fruit, the outer one splitting at maturity into strips (Fig. 49, *B*). Another pretty and common form is the little birds'-nest fungus (*Cyathus*), growing on rotten wood or soil containing much decaying vegetable matter (Fig. 50).

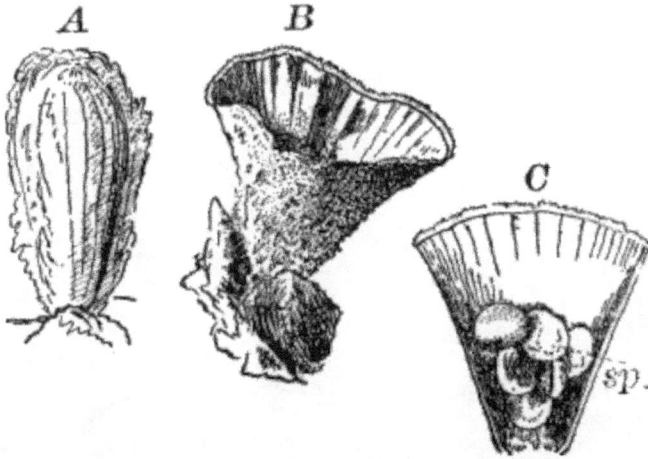

FIG. 50.—Birds'-nest fungus (Cyathus). A, young. B, full grown. C, section through B, showing the "sporangia" (sp.). All twice the natural size.

In the second order the spores are at first protected, as we have seen in *Coprinus*, which belongs to this order, but finally become exposed. Here belong the toadstools and mushrooms (Fig. 51, *B*), the large shelf-shaped fungi (*Polyporus*), so common on tree trunks and rotten logs (Fig. 51, *C*, *D*, *E*), and the prickly fungus (*Hydnum*) (Fig. 51, *G*).

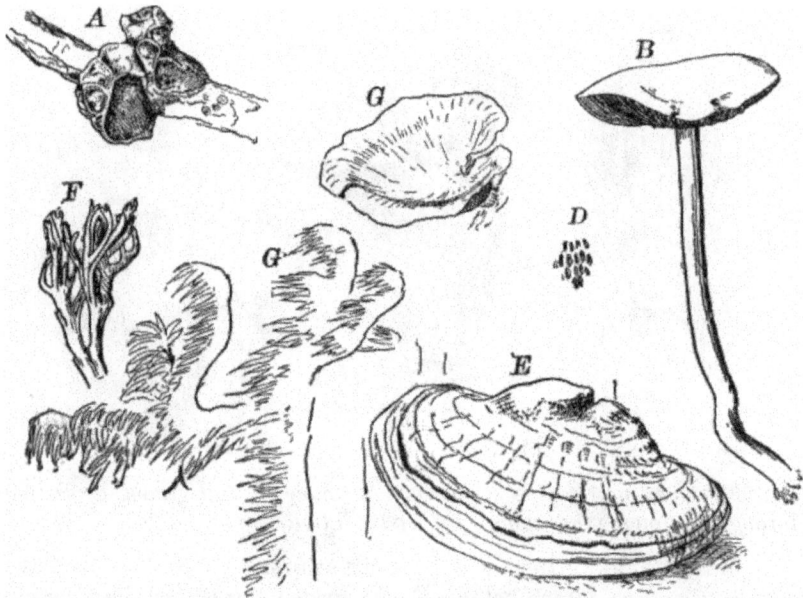

FIG. 51.—*Forms of Basidiomycetes. A, Tremella, one-half natural size. B, Agaricus, natural size. C, E, Polyporus: C, × ½; E, × ¼. D, part of the under surface of D, natural size. F, Clavaria, a small piece, natural size. G, Hydnum, a piece of the natural size.*

Of the last, or naked-fruited forms, the commonest belong to the genus *Clavaria* (Fig. 51, *F*), smooth-branching forms, usually of a brownish color, bearing the spores directly upon the surface of the branches.

CHAPTER XI.
SUB-KINGDOM IV.
Bryophyta.

THE Bryophytes, or mosses, are for the most part land plants, though a few are aquatic, and with very few exceptions are richly supplied with chlorophyll. They are for the most part small plants, few of them being over a few centimetres in height; but, nevertheless, compared with the plants that we have heretofore studied, quite complex in their structure. The lowest members of the group are flattened, creeping plants, or a few of them floating aquatics, without distinct stem and leaves; but the higher ones have a pretty well-developed central axis or stem, with simple leaves attached.

There are two classes—I. Liverworts (*Hepaticæ*), and II. Mosses (*Musci*).

Class I.—The Liverworts.

One of the commonest of this class, and to be had at any time, is named *Madotheca*. It is one of the highest of the class, having distinct stem and leaves. It grows most commonly on the shady side of tree trunks, being most luxuriant near the ground, where the supply of moisture is most constant. It also occurs on stones and rocks in moist places. It closely resembles a true moss in general appearance, and from the scale-like arrangement of its leaves is sometimes called "scale moss."

The leaves (Fig. 52, *A, B*) are rounded in outline unequally, two-lobed, and arranged in two rows on the upper side of the stem, so closely overlapping as to conceal it entirely. On the under side are similar but smaller leaves, less regularly disposed. The stems branch at intervals, the branches spreading out laterally so that the whole plant is decidedly flattened. On the under side are fine, whitish hairs, that fasten it to the substratum. If we examine a number of specimens, especially early in the spring, a difference will be observed in the plants. Some of them will be found to bear peculiar structures (Fig. 52, *C, D*), in which the spores are produced. These are called "sporogonia." They are at first globular, but when ripe open by means of four valves, and discharge a greenish brown mass of spores. An examination of the younger parts of the same plants will probably show small buds (Fig. 54, *H*), which contain the female reproductive organs, from which the sporogonia arise.

FIG. 52.—*A, part of a plant of a leafy liverwort (Madotheca), × 2. B, part of the same, seen from below, × 4. C, a branch with two open sporogonia (sp.), × 4. D, a single sporogonium, × 8.*

On other plants may be found numerous short side branches (Fig. 53, *B*), with very closely set leaves. If these are carefully separated, the antheridia can just be seen as minute whitish globules, barely visible to the naked eye. Plants that, like this one, have the male and female reproductive organs on distinct plants, are said to be "diœcious."

A microscopical examination of the stem and leaves shows their structure to be very simple. The former is cylindrical, and composed of nearly uniform elongated cells, with straight cross-walls. The leaves consist of a single layer of small, roundish cells, which, like those of the stem, contain numerous rounded chloroplasts, to which is due their dark green color.

The tissues are developed from a single apical cell, but it is difficult to obtain good sections through it.

The antheridia are borne singly at the bases of the leaves on the special branches already described (Fig. 53, *A, an.*). By carefully dis-

101

secting with needles such a branch in a drop of water, some of the antheridia will usually be detached uninjured, and may be readily studied, the full-grown ones being just large enough to be seen with the naked eye. They are globular bodies, attached by a stalk composed of two rows of cells. The globular portion consists of a wall of chlorophyll-bearing cells, composed of two layers below, but single above (Fig. 53, C). Within is a mass of excessively small cells, each of which contains a spermatozoid. In the young antheridium (A, an.) the wall is single throughout, and the central cells few in number. To study them in their natural position, thin longitudinal sections of the antheridial branch should be made.

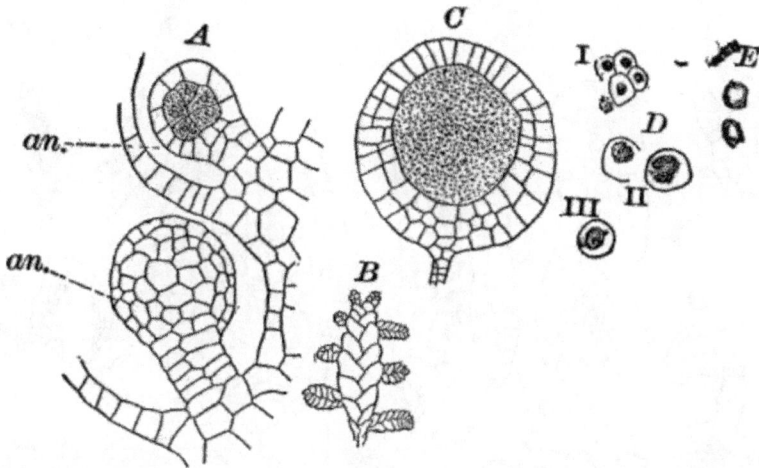

FIG. 53.—A, end of a branch from a male plant of Madotheca. The small side branchlets bear the antheridia, × 2. B, two young antheridia (an.), the upper one seen in optical section, the lower one from without, × 150. C, a ripe antheridium, optical section, × 50. D, sperm cells with young spermatozoids. E, ripe spermatozoids, × 600.

When ripe, if brought into water, the antheridium bursts at the top into a number of irregular lobes that curl back and allow the mass of sperm cells to escape. The spermatozoids, which are derived principally from the nucleus of the sperm cells (53, D) are so small as to make a satisfactory examination possible only with very powerful lenses. The ripe spermatozoid is coiled in a flat spiral (53, E), and has two excessively delicate cilia, visible only under the most favorable circumstances.

The female organ in the bryophytes is called an "archegonium," and differs considerably from anything we have yet studied, but recalls somewhat the structure of the oögonium of *Chara*. They are found in groups, contained in little bud-like branches (54, *H*). In order to study them, a plant should be chosen that has numbers of such buds, and the smallest that can be found should be used. Those containing the young archegonia are very small; but after one has been fertilized, the leaves enclosing it grow much larger, and the bud becomes quite conspicuous, being surrounded by two or three comparatively large leaves. By dissecting the young buds, archegonia in all stages of growth may be found.

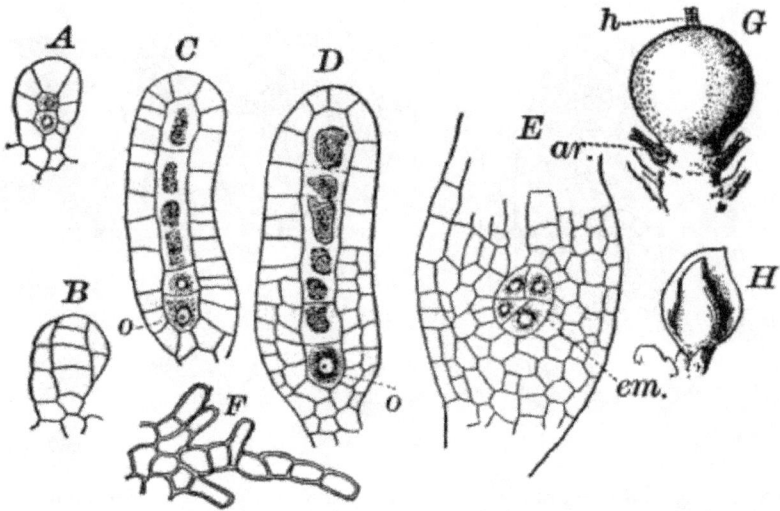

FIG. 54.—*A–D, development of the archegonium of Madotheca. B, surface view, the others in optical section. o, egg cell, × 150. E, base of a fertilized archegonium, containing a young embryo (em.), × 150. F, margin of one of the leaves surrounding the archegonia. G, young sporogonium still surrounded by the much enlarged base of the archegonium. h, neck of the archegonium. ar. abortive archegonia, × 12. H, short branch containing the young sporogonium, × 4.*

When very young the archegonium is composed of an axial row of three cells, surrounded by a single outer layer of cells, the upper ones forming five or six regular rows, which are somewhat twisted (Fig. 54, *A, B*). As it becomes older, the lower part enlarges slightly,

103

the whole looking something like a long-necked flask (C, D). The centre of the neck is occupied by a single row of cells (canal cells), with more granular contents than the outer cells, the lowest cell of the row being somewhat larger than the others (Fig. 54, C, o). When nearly ripe, the division walls of the canal cells are absorbed, and the protoplasm of the lowest cell contracts and forms a globular naked cell, the egg cell (D, o). If a ripe archegonium is placed in water, it soon opens at the top, and the contents of the canal cells are forced out, leaving a clear channel down to the egg cell. If the latter is not fertilized, the inner walls of the neck cells turn brown, and the egg cell dies; but if a spermatozoid penetrates to the egg cell, the latter develops a wall and begins to grow, forming the embryo or young sporogonium.

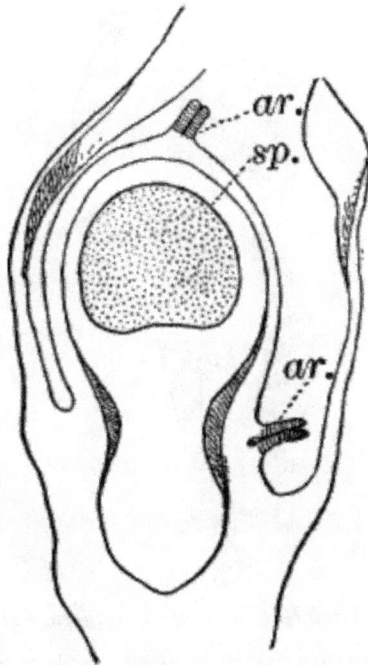

FIG. 55.—*Longitudinal section of a nearly full-grown sporogonium of Madotheca, which has not, however, broken through the overlying cells, × 25. sp. cavity in which the spores are formed. ar. abortive archegonium.*

The first division wall to be formed in the embryo is transverse, and is followed by vertical ones (Fig. 54, E, em.). As the embryo enlarges, the walls of the basal part of the archegonium grow rapidly, so that the embryo remains enclosed in the archegonium until it is nearly full-grown (Fig. 55). As it increases in size, it becomes differentiated into three parts: a wedge-shaped base or "foot" penetrating downward into the upper part of the plant, and serving to supply the embryo with nourishment; second, a stalk supporting the third part, the capsule or spore-bearing portion of the fruit. The capsule is further differentiated into a wall, which later becomes dark colored, and a central cavity, in which are developed special cells, some of which by further division into four parts produce the spores, while the others, elongating enormously, give rise to special cells, called elaters (Fig. 56, B).

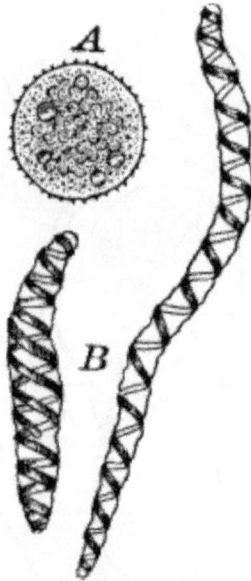

FIG. 56. — *Spore (A) and two elaters (B) of Madotheca, × 300.*

The ripe spores are nearly globular, contain chlorophyll and drops of oil, and the outer wall is brown and covered with fine points (Fig. 56, A). The elaters are long-pointed cells, having on the inner surface of the wall a single or double dark brown spiral band. These bands are susceptible to changes in moisture, and by their movements probably assist in scattering the spores after the sporogonium opens.

Just before the spores are ripe, the stalk of the sporogonium elongates rapidly, carrying up the capsule, which breaks through the archegonium wall, and finally splits into four valves, and discharges the spores.

There are four orders of the liverworts represented in the United States, three of which differ from the one we have studied in being flattened plants, without distinct stems and leaves,—at least, the leaves when present are reduced to little scales upon the lower surface.

The first order (*Ricciaceæ*) are small aquatic forms, or grow on damp ground or rotten logs. They are not common forms, and not likely to be encountered by the student. One of the floating species is shown in figure 57, *A*.

The second order, the horned liverworts (*Anthoceroteæ*), are sometimes to be met with in late summer and autumn, forms growing mostly on damp ground, and at once recognizable by their long-pointed sporogonia, which open when ripe by two valves, like a bean pod (Fig. 57, *B*).

The third order (*Marchantiaceæ*) includes the most conspicuous members of the whole class. Some of them, like the common liverwort (*Marchantia*), shown in Figure 57, *F*, *K*, and the giant liverwort (Fig. 57, *D*), are large and common forms, growing on the ground in shady places, the former being often found also in greenhouses. They are fastened to the ground by numerous fine, silky hairs, and the tissues are well differentiated, the upper surface of the plant having a well-marked epidermis, with peculiar breathing pores, large enough to be seen with the naked eye (Fig. 57, *E*, *J*, *K*) Each of these is situated in the centre of a little area (Fig. 57, *E*), and beneath it is a large air space, into which the chlorophyll-bearing cells (*cl.*) of the plant project (*J*).

The sexual organs are often produced in these forms upon special branches (*G*), or the antheridia may be sunk in discs on the upper side of the stem (*D*, *an.*).

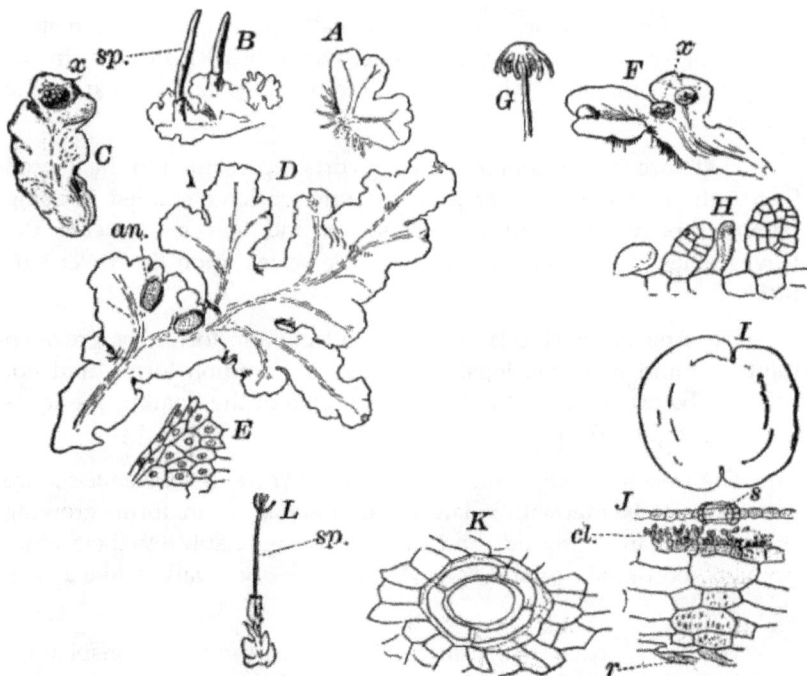

FIG. 57.—*Forms of liverworts. A, Riccia, natural size. B, Anthoceros (horned liverwort), natural size. sp. sporogonia. C, Lunularia, natural size, x, buds. D, giant liverwort (Conocephalus), natural size. an. antheridial disc. E, small piece of the epidermis, showing the breathing pores, × 2. F, common liverwort (Marchantia), × 2. x, cups containing buds. G, archegonial branch of common liverwort, natural size. H, two young buds from the common liverwort, × 150. I, a full-grown bud, × 25. J, vertical section through the body of Marchantia, cutting through a breathing pore (s), × 50. K, surface view of a breathing pore, × 150. L, a leafy liverwort (Jungermannia). sp. sporogonium, × 2.*

Some forms, like *Marchantia* and *Lunularia* (Fig. 57, C), produce little cups (*x*), circular in the first, semicircular in the second, in which special buds (*H, I*) are formed that fall off and produce new plants.

The highest of the liverworts (*Jungermanniaceæ*) are, for the most part, leafy forms like *Madotheca*, and represented by a great many common forms, growing usually on tree trunks, etc. They are much like *Madotheca* in general appearance, but usually very small and inconspicuous, so as to be easily overlooked, especially as their color is apt to be brownish, and not unlike that of the bark on which they grow (Fig. 57, *L*).

107

Class II. — The True Mosses.

The true mosses (*Musci*) resemble in many respects the higher liverworts, such as *Madotheca* or *Jungermannia*, all of them having well-marked stems and leaves. The spore fruit is more highly developed than in the liverworts, but never contains elaters.

A good idea of the general structure of the higher mosses may be had from a study of almost any common species. One of the most convenient, as well as common, forms (*Funaria*) is to be had almost the year round, and fruits at almost all seasons, except midwinter. It grows in close patches on the ground in fields, at the bases of walls, sometimes in the crevices between the bricks of sidewalks, etc. If fruiting, it may be recognized by the nodding capsule on a long stalk, that is often more or less twisted, being sensitive to changes in the moisture of the atmosphere. The plant (Fig. 58, *A*, *B*) has a short stem, thickly set with relatively large leaves. These are oblong and pointed, and the centre is traversed by a delicate midrib. The base of the stem is attached to the ground by numerous fine brown hairs.

The mature capsule is broadly oval in form (Fig. 58, *C*), and provided with a lid that falls off when the spores are ripe. While the capsule is young it is covered by a pointed membranous cap (*B*, *cal.*) that finally falls off. When the lid is removed, a fine fringe is seen surrounding the opening of the capsule, and serving the same purpose as the elaters of the liverworts (Fig. 58, *E*).

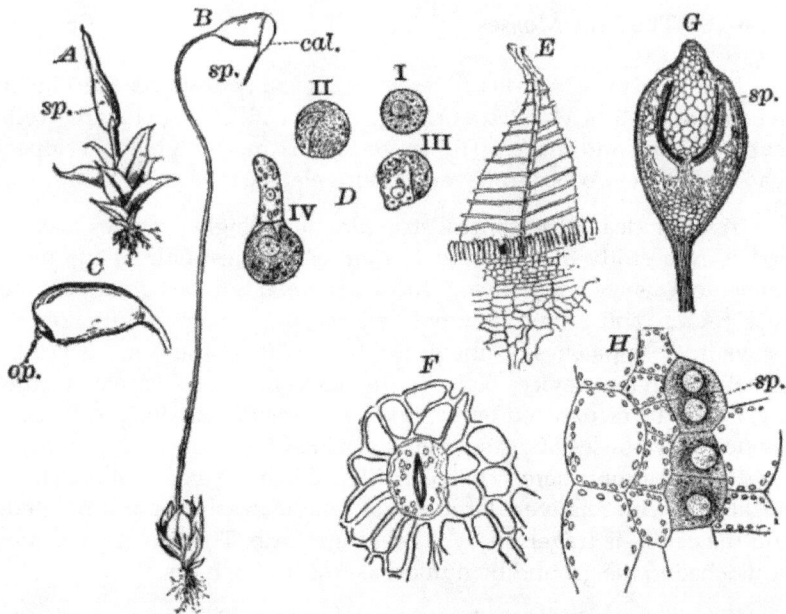

FIG. 58. — *A, fruiting plant of a moss (Funaria), with young sporogonium (sp.),* × 4. *B, plant with ripe sporogonium. cal. calyptra,* × 2. *C, sporogonium with calyptra removed. op. lid,* × 4. *D, spores: I, ungerminated; II–IV, germinating,* × 300. *E, two teeth from the margin of the capsule,* × 50. *F, epidermal cells and breathing pore from the surface of the sporogonium,* × 150. *G, longitudinal section of a young sporogonium,* × 12. *sp. spore mother cells. H, a small portion of G, magnified about 300 times. sp. spore mother cells.*

If the lower part of the stem is carefully examined with a lens, we may detect a number of fine green filaments growing from it, looking like the root hairs, except for their color. Sometimes the ground about young patches of the moss is quite covered by a fine film of such threads, and looking carefully over it probably very small moss plants may be seen growing up here and there from it.

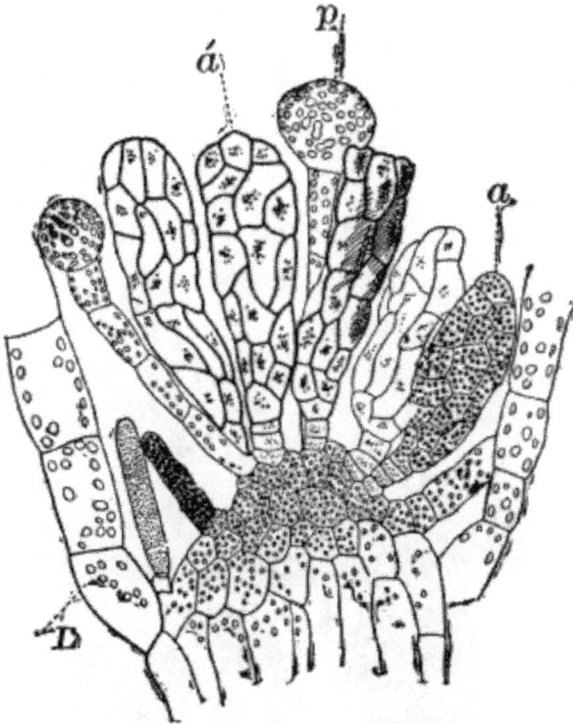

FIG. 59.—*Longitudinal section through the summit of a small male plant of Funaria. a, a□, antheridia. p, paraphysis. L, section of a leaf, × 150.*

This moss is diœcious. The male plants are smaller than the female, and may be recognized by the bright red antheridia which are formed at the end of the stem in considerable numbers, and surrounded by a circle of leaves so that the whole looks something like a flower. (This is still more evident in some other mosses. See Figure 65, E, F.)

The leaves when magnified are seen to be composed of a single layer of cells, except the midrib, which is made up of several thicknesses of elongated cells. Where the leaf is one cell thick, the cells are oblong in form, becoming narrower as they approach the midrib and the margin. They contain numerous chloroplasts imbedded in the layer of protoplasm that lines the wall. The nucleus (Fig. 63, C, n) may usually be seen without difficulty, especially if the leaf is treated with iodine. This plant is one of the best for studying the division of the chloroplasts, which may usually be found in all stages of division (Fig. 63, D). In the chloroplasts, especially if the plant has been exposed to light for several hours, will be found numerous small granules, that assume a bluish tint on the application of iodine, showing them to be starch grains. If the plant is kept in the dark for a day or two, these will be absent, having been used up; but if exposed to the light again, new ones will be formed, showing that they are formed only under the action of light.

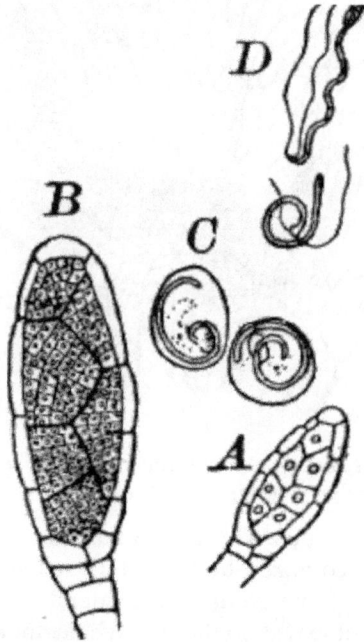

FIG. 60. — A, B, young antheridia of Funaria, optical section, × 150. C, two sperm cells of Atrichum. D, spermatozoids of Sphagnum, × 600.

Starch is composed of carbon, hydrogen, and oxygen, and so far as is known is only produced by chlorophyll-bearing cells, under the influence of light. The carbon used in the manufacture of starch is taken from the atmosphere in the form of carbonic acid, so that green plants serve to purify the atmosphere by the removal of this substance, which is deleterious to animal life, while at the same time the carbon, an essential part of all living matter, is combined in such form as to make it available for the food of other organisms.

111

The marginal cells of the leaf are narrow, and some of them prolonged into teeth.

A cross-section of the stem (63, E) shows on the outside a single row of epidermal cells, then larger chlorophyll-bearing cells, and in the centre a group of very delicate, small, colorless cells, which in longitudinal section are seen to be elongated, and similar to those forming the midrib of the leaf. These cells probably serve for conducting fluids, much as the similar but more perfectly developed bundles of cells (fibro-vascular bundles) found in the stems and leaves of the higher plants.

The root hairs, fastening the plant to the ground, are rows of cells with brown walls and oblique partitions. They often merge insensibly into the green filaments (protonema) already noticed. These latter have usually colorless walls, and more numerous chloroplasts, looking very much like a delicate specimen of *Cladophora* or some similar alga. If a sufficient number of these filaments is examined, some of them will probably show young moss plants growing from them (Fig. 63, A, k), and with a little patience the leafy plant can be traced back to a little bud originating as a branch of the filament. Its diameter is at first scarcely greater than that of the filament, but a series of walls, close together, are formed, so placed as to cut off a pyramidal cell at the top, forming the apical cell of the young moss plant. This apical cell has the form of a three-sided pyramid with the base upward. From it are developed three series of cells, cut off in succession from the three sides, and from these cells are derived all the tissues of the plant which soon becomes of sufficient size to be easily recognizable.

The protonemal filaments may be made to grow from almost any part of the plant by keeping it moist, but grow most abundantly from the base of the stem.

The sexual organs are much like those of the liverworts and are borne at the apex of the stems.

The antheridia (Figs. 59, 60) are club-shaped bodies with a short stalk. The upper part consists of a single layer of large chlorophyll-bearing cells, enclosing a mass of very small, nearly cubical, colorless, sperm cells each of which contains an excessively small spermatozoid.

The young antheridium has an apical cell giving rise to two series of segments (Fig. 60, *A*), which in the earlier stages are very plainly marked.

When ripe the chlorophyll in the outer cells changes color, becoming red, and if a few such antheridia from a plant that has been kept rather dry for a day or two, are teased out in a drop of water, they will quickly open at the apex, the whole mass of sperm cells being discharged at once.

Among the antheridia are borne peculiar hairs (Fig. 59, *p*) tipped by a large globular cell.

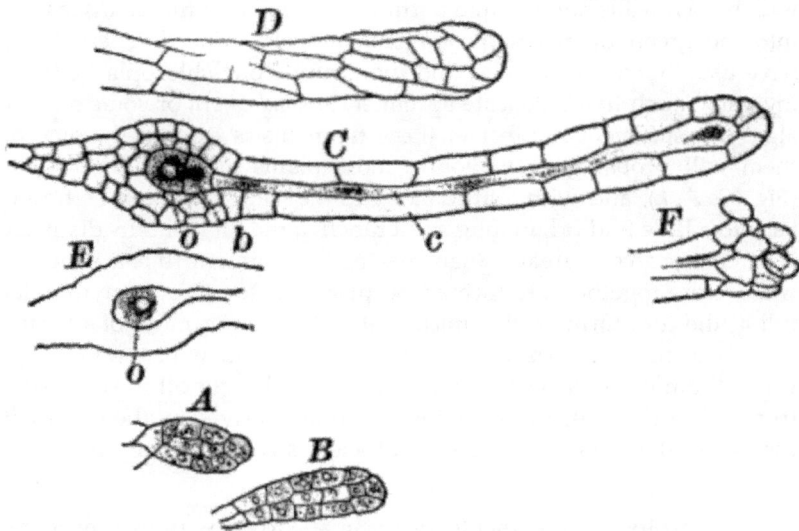

FIG. 61. — *A, B, young; C, nearly ripe archegonium of Funaria, optical section,* × *150. D, upper part of the neck of C, seen from without, showing how it is twisted. E, base of a ripe archegonium. F, open apex of the same,* × *150. o, egg cell. b, ventral canal cell.*

Owing to their small size the spermatozoids are difficult to see satisfactorily and other mosses (*e.g.* peat mosses, Figure 64, the hairy cap moss, Figure 65, *I*), are preferable where obtainable. The spermatozoids of a peat moss are shown in Figure 60, *D*. Like all of the bryophytes they have but two cilia.

The archegonia (Fig. 61) should be looked for in the younger plants in the neighborhood of those that bear capsules. Like the antheridia they occur in groups. They closely resemble those of the liverworts, but the neck is longer and twisted and the base more massive. Usually but a single one of the group is fertilized.

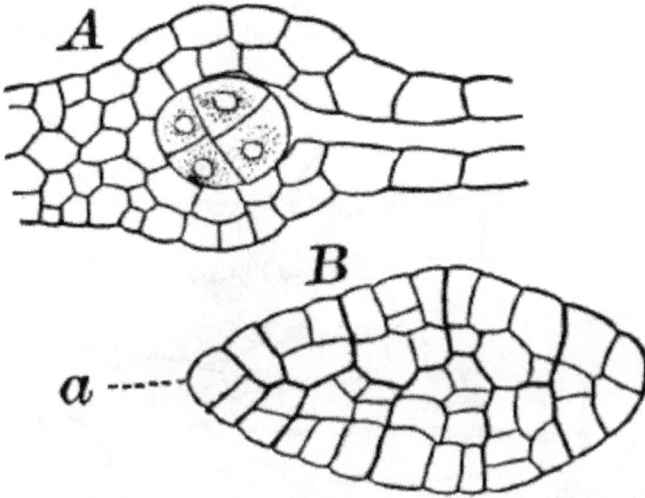

FIG. 62.—A, young embryo of Funaria, still enclosed within the base of the archegonium, × 300. B, an older embryo freed from the archegonium, × 150. a, the apical cell.

To study the first division of the embryo, it is usually necessary to render the archegonium transparent, which may be done by using a little caustic potash; or letting it lie for a few hours in dilute glycerine will sometimes suffice. If potash is used it must be thoroughly washed away, by drawing pure water under the cover glass with a bit of blotting paper, until every trace of the potash is removed. The first wall in the embryo is nearly at right angles to the axis of the archegonium and divides the egg cell into nearly equal parts. This is followed by nearly vertical walls in each cell (Fig. 62, A). Very soon a two-sided apical cell (Fig. 62, B, a) is formed in the upper half of the embryo, which persists until the embryo has reached a considerable size. As in the liverworts the young embryo is completely covered by the growing archegonium wall.

The embryo may be readily removed from the archegonium by adding a little potash to the water in which it is lying, allowing it to remain for a few moments and pressing gently upon the cover glass with a needle. In this way it can be easily forced out of the archegonium, and then by thoroughly washing away the potash, neutralizing if necessary with a little acetic acid, very beautiful preparations may be made. If desired, these may be mounted permanently in glycerine which, however, must be added very gradually to avoid shrinking the cells.

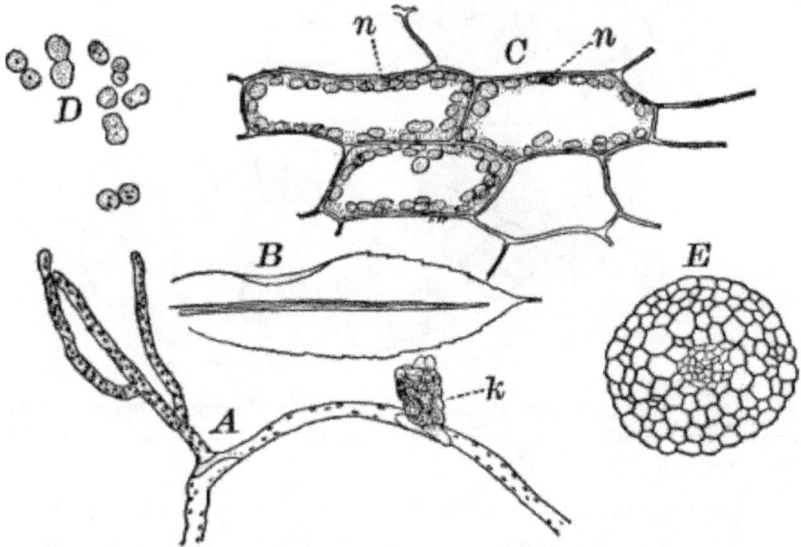

FIG. 63.—A, protonema of Funaria, with a bud (k), × 50. B, outline of a leaf, showing also the thickened midrib, × 12. C, cells of the leaf, × 300. n, nucleus. D, chlorophyll granules undergoing division, × 300. E, cross-section of the stem, × 50.

For some time the embryo has a nearly cylindrical form, but as it approaches maturity the differentiation into stalk and capsule becomes apparent. The latter increases rapidly in diameter, assuming gradually the oval shape of the full-grown capsule. A longitudinal section of the nearly ripe capsule (Fig. 58, G) shows two distinct portions; an outer wall of two layers of cells, and an inner mass of cells in some of which the spores are produced. This inner mass of cells is continuous with the upper part of the capsule, but connected with the

115

side walls and bottom by means of slender, branching filaments of chlorophyll-bearing cells.

The spores arise from a single layer of cells near the outside of the inner mass of cells (*G, sp.*). These cells (*H, sp.*) are filled with glistening, granular protoplasm; have a large and distinct nucleus, and no chlorophyll. They finally become entirely separated and each one gives rise to four spores which closely resemble those of the liverworts but are smaller.

Near the base of the capsule, on the outside, are formed breathing pores (Fig. 58, *F*) quite similar to those of the higher plants.

If the spores are kept in water for a few days they will germinate, bursting the outer brown coat, and the contents protruding through the opening surrounded by the colorless inner spore membrane. The protuberance grows rapidly in length and soon becomes separated from the body of the spore by a wall, and lengthening, more and more, gives rise to a green filament like those we found attached to the base of the full-grown plant, and like those giving rise to buds that develop into leafy plants.

Classification of the Mosses.

The mosses may be divided into four orders: I. The peat mosses (*Sphagnaceæ*); II. *Andreæaceæ*; III. *Phascaceæ*; IV. The common mosses (*Bryaceæ*).

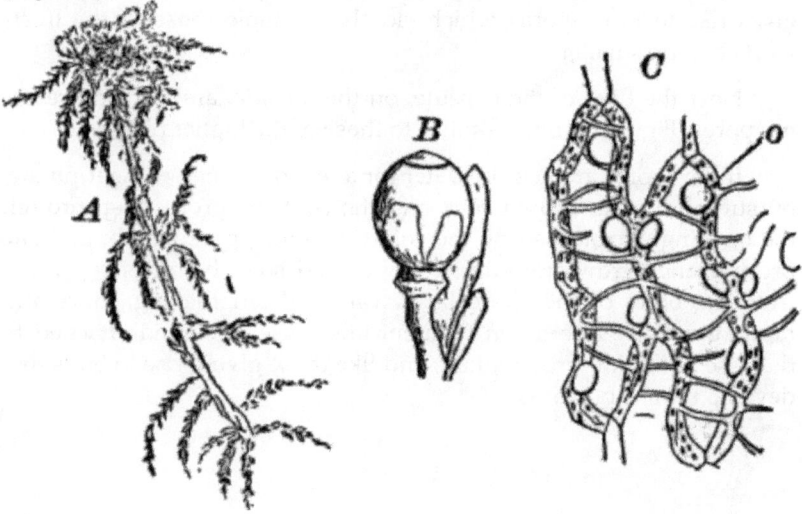

FIG. 64. — *A, a peat moss (Sphagnum), × ½. B, a sporogonium of the same, × 3. C, a portion of a leaf, × 150. The narrow, chlorophyll-bearing cells form meshes, enclosing the large, colorless empty cells, whose walls are marked with thickened bars, and contain round openings (o).*

The peat mosses (Fig. 64) are large pale-green mosses, growing often in enormous masses, forming the foundation of peat-bogs. They are of a peculiar spongy texture, very light when dry, and capable of absorbing a great amount of water. They branch (Fig. 64, *A*), the branches being closely crowded at the top, where the stems continue to grow, dying away below.

117

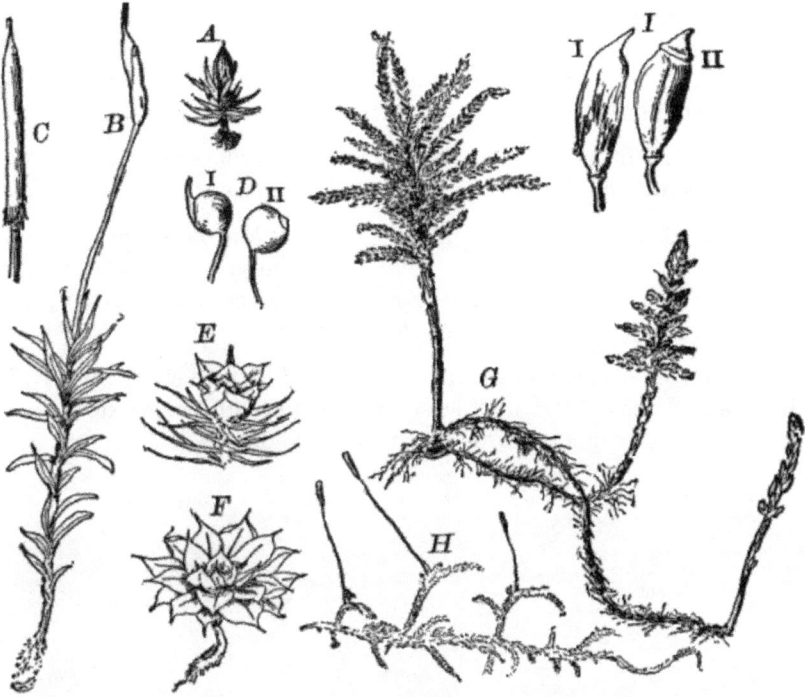

FIG. 65.—*Forms of mosses. A, plant of Phascum,* × 3. *B, fruiting plant of Atrichum,* × 2. *C, young capsule of hairy-cap moss (Polytrichum), covered by the large, hairy calyptra. D, capsules of Bartramia: i, with; ii, without the calyptra. E, upper part of a male plant of Atrichum, showing the flower,* × 2. *F, a male plant of Mnium,* × 4. *G, pine-tree moss (Clemacium),* × 1. *H, Hypnum,* × 1. *I, ripe capsules of hairy-cap moss: i, with; ii, without calyptra.*

The sexual organs are rarely met with, but should be looked for late in autumn or early spring. The antheridial branches are often bright-colored, red or yellow, so as to be very conspicuous. The capsules, which are not often found, are larger than in most of the common mosses, and quite destitute of a stalk, the apparent stalk being a prolongation of the axis of the plant in the top of which the base of the sporogonium is imbedded. The capsule is nearly globular, opening by a lid at the top (Fig. 64, *B*).

A microscopical examination of the leaves, which are quite destitute of a midrib, shows them to be composed of a network of narrow chlorophyll-bearing cells surrounding much larger empty ones whose

118

walls are marked with transverse thickenings, and perforated here and there with large, round holes (Fig. 64, C). It is to the presence of these empty cells that the plant owes its peculiar spongy texture, the growing plants being fairly saturated with water.

The *Andreæaceæ* are very small, and not at all common. The capsule splits into four valves, something like a liverwort.

The *Phascaceæ* are small mosses growing on the ground or low down on the trunks of trees, etc. They differ principally from the common mosses in having the capsule open irregularly and not by a lid. The commonest forms belong to the genus *Phascum* (Fig. 65, A).

The vast majority of the mosses the student is likely to meet with belong to the last order, and agree in the main with the one described. Some of the commoner forms are shown in Figure 65.

CHAPTER XII.
SUB-KINGDOM V.
Pteridophytes.

IF we compare the structure of the sporogonium of a moss or liverwort with the plant bearing the sexual organs, we find that its tissues are better differentiated, and that it is on the whole a more complex structure than the plant that bears it. It, however, remains attached to the parent plant, deriving its nourishment in part through the "foot" by means of which it is attached to the plant.

In the Pteridophytes, however, we find that the sporogonium becomes very much more developed, and finally becomes entirely detached from the sexual plant, developing in most cases roots that fasten it to the ground, after which it may live for many years, and reach a very large size.

The sexual plant, which is here called the "prothallium," is of very simple structure, resembling the lower liverworts usually, and never reaches more than about a centimetre in diameter, and is often much smaller than this.

The common ferns are the types of the sub-kingdom, and a careful study of any of these will illustrate the principal peculiarities of the group. The whole plant, as we know it, is really nothing but the sporogonium, originating from the egg cell in exactly the same way as the moss sporogonium, and like it gives rise to spores which are formed upon the leaves.

The spores may be collected by placing the spore-bearing leaves on sheets of paper and letting them dry, when the ripe spores will be discharged covering the paper as a fine, brown powder. If these are sown on fine, rather closely packed earth, and kept moist and covered with glass so as to prevent evaporation, within a week or two a fine, green, moss-like growth will make its appearance, and by the end of five or six weeks, if the weather is warm, little, flat, heart-shaped plants of a dark-green color may be seen. These look like small liverworts, and are the sexual plants (prothallia) of our ferns (Fig. 66, F). Removing one of these carefully, we find on the lower side numerous fine hairs like those on the lower surface of the liverworts, which fasten it firmly to the ground. By and by, if our culture has been successful, we may find attached to some of the larger of these, little fern plants growing from the under side of the prothallia, and attached to the ground by a delicate root. As the little plant becomes larger the prothallium dies, leaving it attached to the ground as an independent plant, which after a time bears the spores.

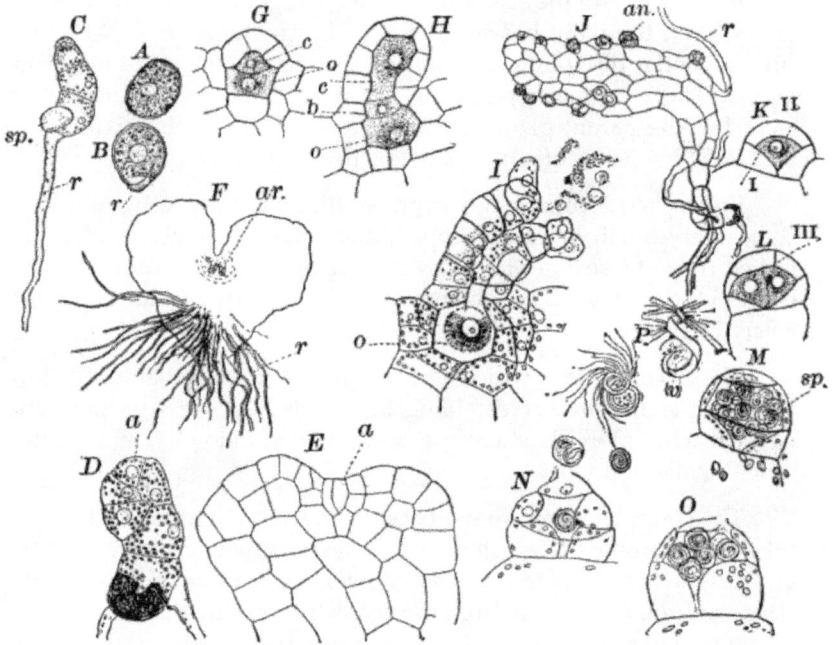

FIG. 66. — *A, spore of the ostrich fern (Onoclea), with the outer coat removed. B, germinating spore, × 150. C, young prothallium, × 50. r, root hair. sp. spore membrane. D, E, older prothallia. a, apical cell, × 150. F, a female prothallium, seen from below, × 12. ar. archegonia. G, H, young archegonia, in optical section, × 150. o, central cell. b, ventral canal cell. c, upper canal cell. I, a ripe archegonium in the act of opening, × 150. o, egg cell. J, a male prothallium, × 50. an. antheridia. K, L, young antheridia, in optical section, × 300. M, ripe antheridium, × 300. sp. sperm cells. N, O, antheridia that have partially discharged their contents, × 300. P, spermatozoids, killed with iodine, × 500. v, vesicle attached to the hinder end.*

In choosing spores for germination it is best to select those of large size and containing abundant chlorophyll, as they germinate more readily. Especially favorable for this purpose are the spores of the ostrich fern (*Onoclea struthiopteris*) (Fig. 70, *I, J*), or the sensitive fern (*O. sensibilis*). Another common and readily grown species is the lady fern (*Asplenium filixfœmina*) (Fig. 70, *H*). The spores of most ferns

121

retain their vitality for many months, and hence can be kept dry until wanted.

The first stages of germination may be readily seen by sowing the spores in water, where, under favorable circumstances, they will begin to grow within three or four days. The outer, dry, brown coat of the spore is first ruptured, and often completely thrown off by the swelling of the spore contents. Below this is a second colorless membrane which is also ruptured, but remains attached to the spore. Through the orifice in the second coat, the inner delicate membrane protrudes in the form of a nearly colorless papilla which rapidly elongates and becomes separated from the body of the spore by a partition, constituting the first root hair (Fig. 66, B, C, r). The body of the spore containing most of the chlorophyll elongates more slowly, and divides by a series of transverse walls so as to form a short row of cells, resembling in structure some of the simpler algæ (C).

In order to follow the development further, spores must be sown upon earth, as they do not develop normally in water beyond this stage.

In studying plants grown on earth, they should be carefully removed and washed in a drop of water so as to remove, as far as possible, any adherent particles, and then may be mounted in water for microscopic examination.

In most cases, after three or four cross-walls are formed, two walls arise in the end cell so inclined as to enclose a wedge-shaped cell (a) from which are cut off two series of segments by walls directed alternately right and left (Fig. 66, D, E, a), the apical cell growing to its original dimensions after each pair of segments is cut off. The segments divide by vertical walls in various directions so that the young plant rapidly assumes the form of a flat plate of cells attached to the ground by root hairs developed from the lower surfaces of the cells, and sometimes from the marginal ones. As the division walls are all vertical, the plant is nowhere more than one cell thick. The marginal cells of the young segments divide more rapidly than the inner ones, and soon project beyond the apical cell which thus comes to lie at the bottom of a cleft in the front of the plant which in consequence becomes heart-shaped (E, F). Sooner or later the apical cell ceases to form regular segments and becomes indistinguishable from the other cells.

In the ostrich fern and lady fern the plants are diœcious. The male plants (Fig. 66, *J*) are very small, often barely visible to the naked eye, and when growing thickly form dense, moss-like patches. They are variable in form, some irregularly shaped, others simple rows of cells, and some have the heart shape of the larger plants.

The female plants (Fig. 66, *F*) are always comparatively large and regularly heart-shaped, occasionally reaching a diameter of nearly or quite one centimetre, so that they are easily recognizable without microscopical examination.

All the cells of the plant except the root hairs contain large and distinct chloroplasts much like those in the leaves of the moss, and like them usually to be found in process of division.

The archegonia arise from cells of the lower surface, just behind the notch in front (Fig. 66, *F, ar.*). Previous to their formation the cells at this point divide by walls parallel to the surface of the plant, so as to form several layers of cells, and from the lowest layer of cells the archegonia arise. They resemble those of the liverworts but are shorter, and the lower part is completely sunk within the tissues of the plant (Fig. 66, *G, I*). They arise as single surface cells, this first dividing into three by walls parallel to the outer surface. The lower cell undergoes one or two divisions, but undergoes no further change; the second cell (*C, o*), becomes the egg cell, and from it is cut off another cell (*c*), the canal cell of the neck; the uppermost of the three becomes the neck. There are four rows of neck cells, the two forward ones being longer than the others, so that the neck is bent backward. In the full-grown archegonium, there are two canal cells, the lower one (*H, b*) called the ventral canal cell, being smaller than the other.

Shortly before the archegonium opens, the canal cells become disorganized in the same way as in the bryophytes, and the protoplasm of the central cell contracts to form the egg cell which shows a large, central nucleus, and in favorable cases, a clear space at the top called the "receptive spot," as it is here that the spermatozoid enters. When ripe, if placed in water, the neck cells become very much distended and finally open widely at the top, the upper ones not infrequently being detached, and the remains of the neck cells are forced out (Fig. 66, *I*).

The antheridia (Fig. 66. *J, M*) arise as simple hemispherical cells, in which two walls are formed (*K* I, II), the lower funnel-shaped, the

123

upper hemispherical and meeting the lower one so as to enclose a central cell (shaded in the figure), from which the sperm cells arise. Finally, a ring-shaped wall (L III) is formed, cutting off a sort of cap cell, so that the antheridium at this stage consists of a central cell, surrounded by three other cells, the two lower ring-shaped, the upper disc-shaped. The central cell, which contains dense, glistening protoplasm, is destitute of chlorophyll, but the outer cells have a few small chloroplasts. The former divides repeatedly, until a mass of about thirty-two sperm cells is formed, each giving rise to a large spirally-coiled spermatozoid. When ripe, the mass of sperm cells crowds so upon the outer cells as to render them almost invisible, and as they ripen they separate by a partial dissolving of the division walls. When brought into water, the outer cells of the antheridium swell strongly, and the matter derived from the dissolved walls of the sperm cells also absorbs water, so that finally the pressure becomes so great that the wall of the antheridium breaks, and the sperm cells are forced out by the swelling up of the wall cells (N, O). After lying a few moments in the water, the wall of each sperm cell becomes completely dissolved, and the spermatozoids are released, and swim rapidly away with a twisting movement. They may be killed with a little iodine, when each is seen to be a somewhat flattened band, coiled several times. At the forward end, the coils are smaller, and there are numerous very long and delicate cilia. At the hinder end may generally be seen a delicate sac (P, v), containing a few small granules, some of which usually show the reaction of starch, turning blue when iodine is applied.

In studying the development of the antheridia, it is only necessary to mount the plants in water and examine them directly; but the study of the archegonia requires careful longitudinal sections of the prothallium. To make these, the prothallium should be placed between small pieces of pith, and the razor must be very sharp. It may be necessary to use a little potash to make the sections transparent enough to see the structure, but this must be used cautiously on account of the great delicacy of the tissues.

If a plant with ripe archegonia is placed in a drop of water, with the lower surface uppermost, and at the same time male plants are put with it, and the whole covered with a cover glass, the archegonia and antheridia will open simultaneously; and, if examined with the microscope, we shall see the spermatozoids collect about the open archegonia, to which they are attracted by the substance forced out when it opens. With a little patience, one or more may be seen to enter the open neck through which it forces itself, by a slow twisting movement, down to the egg cell. In order to make the experiment successful, the plants should be allowed to become a little dry, care being taken that no water is poured over them for a day or two beforehand.

The first divisions of the fertilized egg cell resemble those in the moss embryo, except that the first wall is parallel with the archegonium axis, instead of at right angles to it. Very soon, however, the embryo becomes very different, four growing points being established instead of the single one found in the moss embryo. The two growing points on the side of the embryo nearest the archegonium neck grow faster than the others, one of these outstripping the other, and soon becoming recognizable as the first leaf of the embryo (Fig. 67, A, L). The other (r) is peculiar, in having its growing point covered by several layers of cells, cut off from its outer face, a peculiarity which we shall find is characteristic of the roots of all the higher plants, and, indeed, this is the first root of the young fern. Of the other two growing points, the one next the leaf grows slowly, forming a blunt cone (st.), and is the apex of the stem. The other (f) has no definite form, and serves merely as an organ of absorption, by means of which nourishment is supplied to the embryo from the prothallium; it is known as the foot.

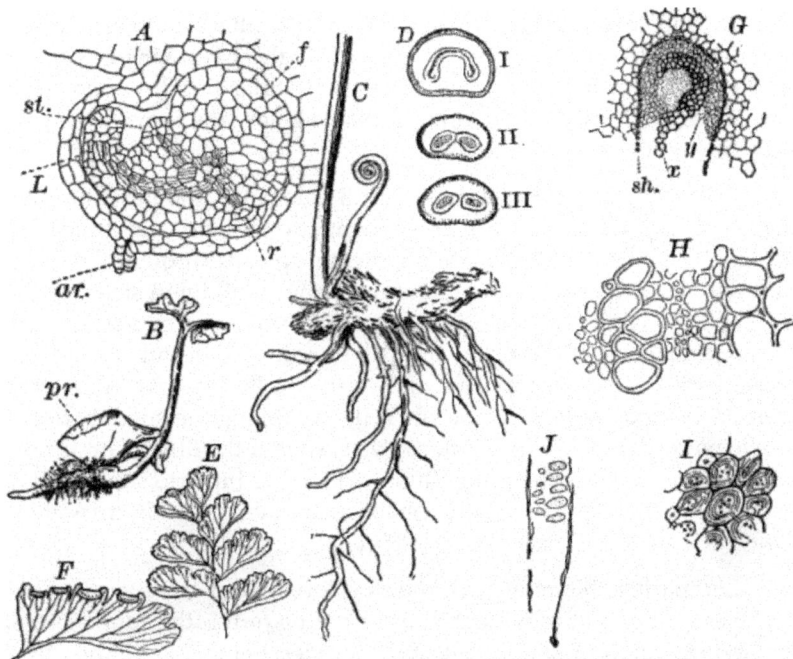

FIG. 67. — A, *embryo of the ostrich fern just before breaking through the prothallium,* × 50. *st. apex of stem. l, first leaf. r, first root. ar. neck of the archegonium. B, young plant, still attached to the prothallium (pr.). C, underground stem of the maiden-hair fern (Adiantum), with one young leaf, and the base of an older one,* × 1. *D, three cross-sections of a leaf stalk:* I, *nearest the base;* III, *nearest the blade of the leaf, showing the division of the fibro-vascular bundle,* × 5. *E, part of the blade of the leaf,* × ½. *F, a single spore-bearing leaflet, showing the edge folded over to cover the sporangia,* × 1. *G, part of the fibro-vascular bundle of the leaf stalk (cross-section),* × 50. *x, woody part of the bundle. y, bast. sh. bundle sheath. H, a small portion of the same bundle,* × 150. *I, stony tissue from the underground stem,* × 150. *J, sieve tube from the underground stem,* × 300.

Up to this point, all the cells of the embryo are much alike, and the embryo, like that of the bryophytes, is completely surrounded by the enlarged base of the archegonium (compare Fig. 67, *A*, with Fig. 55); but before the embryo breaks through the overlying cells a differentiation of the tissues begins. In the axis of each of the four divisions the cells divide lengthwise so as to form a cylindrical mass of narrow cells, not unlike those in the stem of a moss. Here, however, some of the cells undergo a further change; the walls thicken in places,

126

and the cells lose their contents, forming a peculiar conducting tissue (tracheary tissue), found only in the two highest sub-kingdoms. The whole central cylinder is called a "fibro-vascular bundle," and in its perfect form, at least, is found in no plants below the ferns, which are also the first to develop true roots.

The young root and leaf now rapidly elongate, and burst through the overlying cells, the former growing downward and becoming fastened in the ground, the latter growing upward through the notch in the front of the prothallium, and increasing rapidly in size (Fig. 67, B). The leaf is more or less deeply cleft, and traversed by veins which are continuations of the fibro-vascular bundle of the stalk, and themselves fork once or twice. The surface of the leaf is covered with a well-developed epidermis, and the cells occupying the space between the veins contain numerous chloroplasts, so that the little plant is now quite independent of the prothallium, which has hitherto supported it. As soon as the fern is firmly established, the prothallium withers away.

Comparing this now with the development of the sporogonium in the bryophytes, it is evident that the young fern is the equivalent of the sporogonium or spore fruit of the former, being, like it, the direct product of the fertilized egg cell; and the prothallium represents the moss or liverwort, upon which are borne the sexual organs. In the fern, however, the sporogonium becomes entirely independent of the sexual plant, and does not produce spores until it has reached a large size, living many years. The sexual stage, on the other hand, is very much reduced, as we have seen, being so small as to be ordinarily completely overlooked; but its resemblance to the lower liverworts, like *Riccia*, or the horned liverworts, is obvious. The terms oöphyte (egg-bearing plant) and sporophyte (spore-bearing plant, or sporogonium) are sometimes used to distinguish between the sexual plant and the spore-bearing one produced from it.

The common maiden-hair fern (*Adiantum pedatum*) has been selected here for studying the structure of the full-grown sporophyte, but almost any other common fern will answer. The maiden-hair fern is common in rich woods, and may be at once recognized by the form of its leaves. These arise from a creeping, underground stem (Fig. 67, C), which is covered with brownish scales, and each leaf consists of a slender stalk, reddish brown or nearly black in color, which divides into two equal branches at the top. Each of these main branches bears a row of smaller ones on the outside, and these have a row of delicate leaflets on each side (Fig. 67, E). The stem of the plant is fastened to the ground by means of numerous stout roots. The youngest of these, near the growing point of the stem, are unbranched, but the older ones branch extensively (C).

On breaking the stem across, it is seen to be dark-colored, except in the centre, which is traversed by a woody cylinder (fibro-vascular bundle) of a lighter color. This is sometimes circular in sections, sometimes horse-shoe shaped. Where the stem branches, the bundle of the branch may be traced back to where it joins that of the main stem.

A thin cross-section of the stem shows, when magnified, three regions. First, an outer row of cells, often absent in the older portions; this is the epidermis. Second, within the epidermis are several rows of cells similar to the epidermal cells, but somewhat larger, and like them having dark-brown walls. These merge gradually into larger cells, with thicker golden brown walls (Fig. 67, I). The latter, if sufficiently magnified, show distinct striation of the walls, which are often penetrated by deep narrow depressions or "pits." This thick-walled tissue is called "stony tissue" (schlerenchyma). All the cells contain numerous granules, which the iodine test shows to be starch. All of this second region lying between the epidermis and the fibro-vascular bundle is known as the ground tissue. The third region (fibro-vascular) is, as we have seen without the microscope, circular or horse-shoe shaped. It is sharply separated from the ground tissue by a row of small cells, called the "bundle sheath." The cross-section of the bundle of the leaf stalk resembles, almost exactly, that of the stem; and, as it is much easier to cut, it is to be preferred in studying the arrangement of the tissues of the bundle (Fig. 67, G). Within the bundle sheath (*sh.*) there are two well-marked regions, a central band (x) of large empty cells, with somewhat angular outlines, and distinctly separated walls; and an outer portion (y) filling up the space between these central cells and the bundle sheath. The central tissue (x) is called the woody tissue (xylem); the outer, the bast (phloem). The latter is composed of smaller cells of variable form, and with softer walls than the wood cells.

A longitudinal section of either the stem or leaf stalk shows that all the cells are decidedly elongated, especially those of the fibro-vascular bundle. The xylem (Fig. 68, C, x) is made up principally of large empty cells, with pointed ends, whose walls are marked with closely set, narrow, transverse pits, giving them the appearance of little ladders, whence they are called "scalariform," or ladder-shaped markings. These empty cells are known as "tracheids," and tissue composed of such empty cells, "tracheary tissue." Besides the tracheids, there are a few small cells with oblique ends, and with some granular contents.

The phloem is composed of cells similar to the latter, but there may also be found, especially in the stem, other larger ones (Fig. 67, J), whose walls are marked with shallow depressions, whose bottoms are finely pitted. These are the so-called "sieve tubes."

For microscopical examination, either fresh or alcoholic material may be used, the sections being mounted in water. Potash will be found useful in rendering opaque sections transparent.

The leaves, when young, are coiled up (Fig. 67, C), owing to growth in the earlier stages being greater on the lower than on the upper side. As the leaf unfolds, the stalk straightens, and the upper portion (blade) becomes flat.

The general structure of the leaf stalk may be understood by making a series of cross-sections at different heights, and examining them with a hand lens. The arrangement is essentially the same as in the stem. The epidermis and immediately underlying ground tissue are dark-colored, but the inner ground tissue is light-colored, and much softer than the corresponding part of the stem; and some of the outer cells show a greenish color, due to the presence of chlorophyll.

The section of the fibro-vascular bundle differs at different heights. Near the base of the stalk (Fig. D I) it is horseshoe-shaped; but, if examined higher up, it is found to divide (II, III), one part going to each of the main branches of the leaf. These secondary bundles divide further, forming the veins of the leaflets.

The leaflets (E, F) are one-sided, the principal vein running close to the lower edge, and the others branching from it, and forking as they approach the upper margin, which is deeply lobed, the lobes being again divided into teeth. The leaflets are very thin and delicate, with extremely smooth surface, which sheds water perfectly. If the plant is a large one, some of the leaves will probably bear spores. The spore-bearing leaves are at once distinguished by having the middle of each lobe of the leaflets folded over upon the lower side (F). On lifting one of these flaps, numerous little rounded bodies (spore cases) are seen, whitish when young, but becoming brown as they ripen. If a leaf with ripe spore cases is placed upon a piece of paper, as it dries the spores are discharged, covering the paper with the spores, which look like fine brown powder.

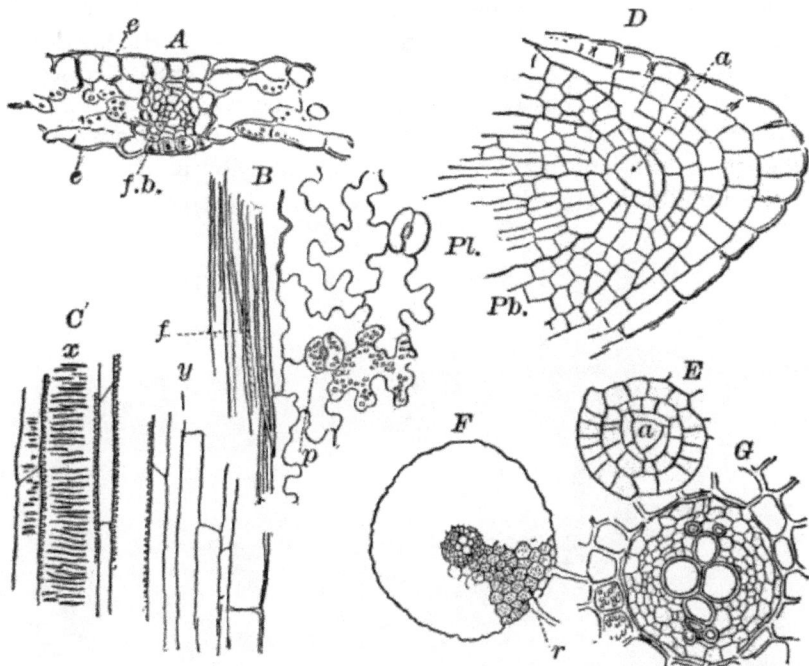

FIG. 68.—A, vertical section of the leaf of the maiden-hair fern, which has cut across a vein (f.b.), × 150. B, surface view of the epidermis from the lower surface of a leaf. f, vein. p, breathing pore, × 150. C, longitudinal section of the fibro-vascular bundle of the leaf stalk, showing tracheids with ladder-shaped markings, × 150. D, longitudinal section through the tip of a root, × 150. a, apical cell. Pl. young fibro-vascular bundle. Pb. young ground tissue. E, cross-section of the root, through the region of the apical cell (a), × 150. F, cross-section through a full-grown root, × 25. r, root hairs. G, the fibro-vascular bundle of the same, × 150.

A microscopical examination of the leaf stalk shows the tissues to be almost exactly like those of the stem, except the inner ground tissue, whose cells are thin-walled and colorless (soft tissue or "parenchyma") instead of stony tissue. The structure of the blade of the leaf, however, shows a number of peculiarities. Stripping off a little of the epidermis with a needle, or shaving off a thin slice with a razor, it may be examined in water, removing the air if necessary with alcohol. It is composed of a single layer of cells, of very irregular outline, except where it overlies a vein (Fig. 68, *B*, *f*). Here the cells are long and narrow, with heavy walls. The epidermal cells contain numerous chloroplasts, and on the under surface of the leaf breathing pores (*stomata*, sing. *stoma*), not unlike those on the capsules of some of the bryophytes. Each breathing pore consists of two special crescent-shaped epidermal cells (guard cells), enclosing a central opening or pore communicating with an air space below. They arise from cells of the young epidermis that divide by a longitudinal wall, that separates in the middle, leaving the space between.

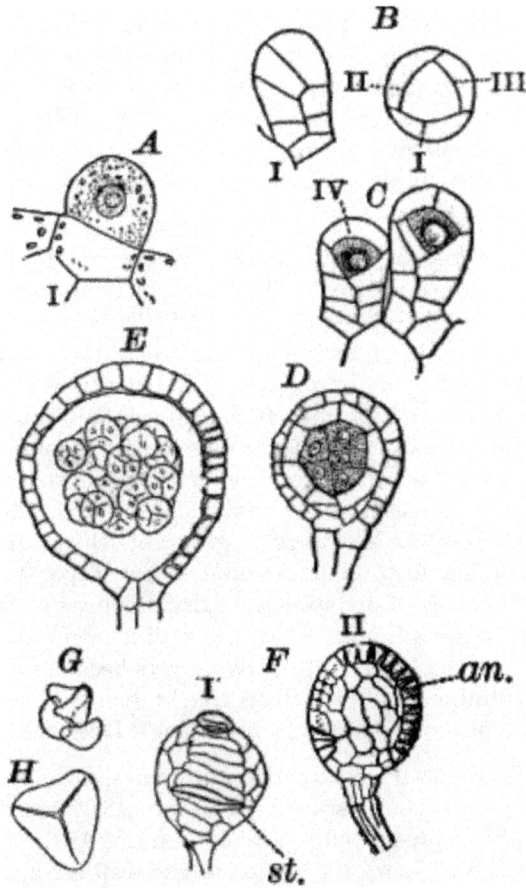

FIG. 69.—A, mother cell of the sporangium of the maiden-hair fern, × 300. B, young sporangium, surface view, × 150: I, from the side; II, from above. C-E, successive stages in the development of the sporangium seen in optical section, × 150. F, nearly ripe sporangium, × 50: I, from in front; II, from the side. an. ring. st. point of opening. G, group of four spores, × 150. H, a single spore, × 300.

By holding a leaflet between two pieces of pith, and using a very sharp razor, cross-sections can be made. Such a section is shown in Fig. 68, A. The epidermis (e) bounds the upper and lower surfaces, and if a vein (f.b.) is cut across its structure is found to be like that of the fibro-vascular bundle of the leaf stalk, but much simplified.

The ground tissue of the leaf is composed of very loose, thin-walled cells, containing numerous chloroplasts. Between them are large and numerous intercellular spaces, filled with air, and communicating with the breathing pores. These are the principal assimilating cells of the plant; *i.e.* they are principally concerned in the absorption and decomposition of carbonic acid from the atmosphere, and the manufacture of starch.

The spore cases, or sporangia (Fig. 69), are at first little papillæ (*A*), arising from the epidermal cells, from which they are early cut off by a cross-wall. In the upper cell several walls next arise, forming a short stalk, composed of three rows of cells, and an upper nearly spherical cell — the sporangium proper. The latter now divides by four walls (*B*, *C*, I–IV), into a central tetrahedral cell, and four outer ones. The central cell, whose contents are much denser than the outer ones, divides again by walls parallel to those first formed, so that the young sporangium now consists of a central cell, surrounded by two outer layers of cells. From the central cell a group of cells is formed by further divisions (*D*), which finally become entirely separated from each other. The outer cells of the spore case divide only by walls, at right angles to their outer surface, so that the wall is never more than two cells thick. Later, the inner of these two layers becomes disorganized, so that the central mass of cells floats free in the cavity of the sporangium, which is now surrounded by but a single layer of cells (*E*).

Each of the central cells divides into four spores, precisely as in the bryophytes. The young spores (*G*, *H*) are nearly colorless and are tetrahedral (like a three-sided pyramid) in form. As they ripen, chlorophyll is formed in them, and some oil. The wall becomes differentiated into three layers, the outer opaque and brown, the two inner more delicate and colorless.

Running around the outside of the ripe spore case is a single row of cells (*an.*), differing from the others in shape, and having their inner walls thickened. Near the bottom, two (sometimes four) of these cells are wider than the others, and their walls are more strongly thickened. It is at this place (*st.*) that the spore case opens. When the ripe sporangium becomes dry, the ring of thickened cells (*an.*) contracts more strongly than the others, and acts like a spring pulling the sporangium open and shaking out the spores, which germinate readily under favorable conditions, and form after a time the sexual plants (prothallia).

The roots of the sporophyte arise in large numbers, the youngest being always nearest the growing point of the stem or larger roots (Fig. 67, C). The growing roots are pointed at the end which is also light-colored, the older parts becoming dark brown. A cross-section of the older portions shows a dark-brown ground tissue with a central, light-colored, circular, fibro-vascular bundle (Fig. 68, F). Growing from its outer surface are numerous brown root hairs (r).

When magnified the walls of all the outer cells (epidermis and ground tissue) are found to be dark-colored but not very thick, and the cells are usually filled with starch. There is a bundle sheath of much-flattened cells separating the fibro-vascular bundle from the ground tissue. The bundle (Fig. 68, G) shows a band of tracheary tissue in the centre surrounded by colorless cells, all about alike.

All of the organs of the fern grow from a definite apical cell, but it is difficult to study except in the root.

Selecting a fresh, pretty large root, a series of thin longitudinal sections should be made either holding the root directly in the fingers or placing it between pieces of pith. In order to avoid drying of the sections, as is indeed true in cutting any delicate tissue, it is a good plan to wet the blade of the razor. If the section has passed through the apex, it will show the structure shown in Figure 68, D. The apical cell (a) is large and distinct, irregularly triangular in outline. It is really a triangular pyramid (tetrahedron) with the base upward, which is shown by making a series of cross-sections through the root tip, and comparing them with the longitudinal sections. The cross-section of the apical cell (Fig. L) appears also triangular, showing all its faces to be triangles. Regular series of segments are cut off in succession from each of the four faces of the apical cell. These segments undergo regular divisions also, so that very early a differentiation of the tissues is evident, and the three tissue systems (epidermal, ground, and fibro-vascular) may be traced almost to the apex of the root (68, D). From the outer series of segments is derived the peculiar structure (root cap) covering the delicate growing point and protecting it from injury.

The apices of the stem and leaves, being otherwise protected, develop segments only from the sides of the apical cell, the outer face never having segments cut off from it.

134

CHAPTER XIII.
CLASSIFICATION OF THE PTERIDOPHYTES.

THERE are three well-marked classes of the Pteridophytes: the ferns (*Filicinæ*); horse-tails (*Equisetinæ*); and the club mosses (*Lycopodinæ*).

Class I. — Ferns (*FILICINÆ*).

The ferns constitute by far the greater number of pteridophytes, and their general structure corresponds with that of the maiden-hair fern described. There are three orders, of which two, the true ferns (*Filices*) and the adder-tongues (*Ophioglossaceæ*), are represented in the United States. A third order, intermediate in some respects between these two, and called the ringless ferns (*Marattiaceæ*), has no representatives within our territory.

The classification is at present based largely upon the characters of the sporophyte, the sexual plants being still very imperfectly known in many forms.

The adder-tongues (*Ophioglossaceæ*) are mostly plants of rather small size, ranging from about ten to fifty centimetres in height. There are two genera in the United States, the true adder-tongues (*Ophioglossum*) and the grape ferns (*Botrychium*). They send up but one leaf each year, and this in fruiting specimens (Fig. 70, *A*) is divided into two portions, the spore bearing (*x*) and the green vegetative part. In *Botrychium* the leaves are more or less deeply divided, and the sporangia distinct (Fig. 71, *B*). In *Ophioglossum* the sterile division of the leaf is usually smooth and undivided, and the spore-bearing division forms a sort of spike, and the sporangia are much less distinct. The sporangia in both differ essentially from those of the true ferns in not being derived from a single epidermal cell, but are developed in part from the ground tissue of the leaf.

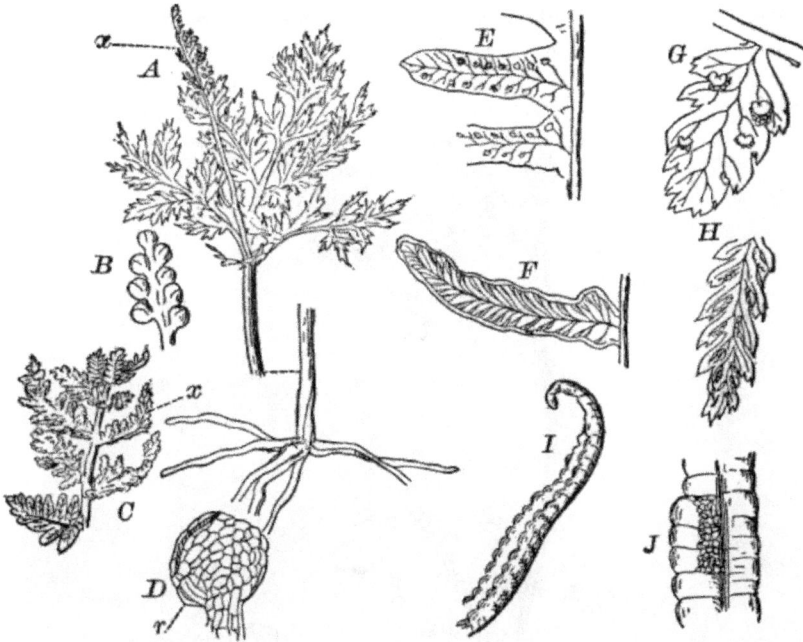

FIG. 70.—*Forms of ferns. A, grape fern (Botrychium), × ½. x, fertile part of the leaf. B, sporangia of Botrychium, × 3. C, flowering fern (Osmunda). x, spore-bearing leaflets, × ½. D, a sporangium of Osmunda, × 25. r, ring. E, Polypodium, × 1. F, brake (Pteris), × 1. G, shield fern (Aspidium), × 2. H, spleen-wort (Asplenium), × 2. I, ostrich fern (Onoclea), × 1. J, the same, with the incurved edges of the leaflet partially raised so as to show the masses of sporangia beneath, × 2.*

In the true ferns (*Filices*), the sporangia resemble those already described, arising in all (unless possibly *Osmunda*) from a single epidermal cell.

One group, the water ferns (*Rhizocarpeæ*), produce two kinds of spores, large and small. The former produce male, the latter female prothallia. In both cases the prothallium is small, and often scarcely protrudes beyond the spore, and may be reduced to a single archegonium or antheridium (Fig. 71, *B, C*) with only one or two cells representing the vegetative cells of the prothallium (*v*). The water ferns are all aquatic or semi-aquatic plants, few in number and scarce or local in their distribution. The commonest are those of the genus *Marsilia*

136

(Fig. 71, *A*), looking like a four-leaved clover. Others (*Salvinia, Azolla*) are floating forms (Fig. 71, *D*).

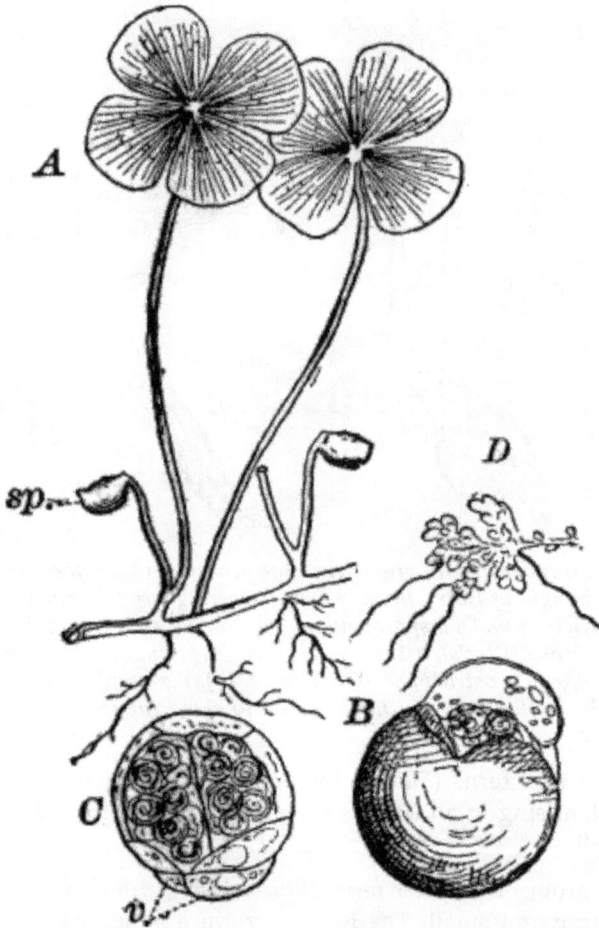

FIG. 71.—*A, Marsilia, one of the Rhizocarpeæ (after Underwood). sp. the "fruits" containing the sporangia. B, a small spore of Pilularia, with the ripe antheridium protruding, × 180. C, male prothallium removed from the spore, × 180. D, Azolla (after Sprague), × 1.*

Of the true ferns there are a number of families distinguished mainly by the position of the sporangia, as well as by some differences in their structure. Of our common ferns, those differing most widely from the types are the flowering ferns (*Osmunda*), shown in Figure 70, *C, D*. In these the sporangia are large and the ring (*r*) rudimentary. The leaflets bearing the sporangia are more or less contracted and covered completely with the sporangia, sometimes all the leaflets of the spore-bearing leaf being thus changed, sometimes only a few of them, as in the species figured.

Our other common ferns have the sporangia in groups (*sori*, sing. *sorus*) on the backs of the leaves. These sori are of different shape in different genera, and are usually protected by a delicate membranous covering (indusium). Illustrations of some of the commonest genera are shown in Figure 70, *E, J*.

Class II. — Horse-tails (*EQUISETINÆ*).

The second class of the pteridophytes includes the horse-tails (*Equisetinæ*) of which all living forms belong to a single genus (*Equisetum*). Formerly they were much more numerous than at present, remains of many different forms being especially abundant in the coal formations.

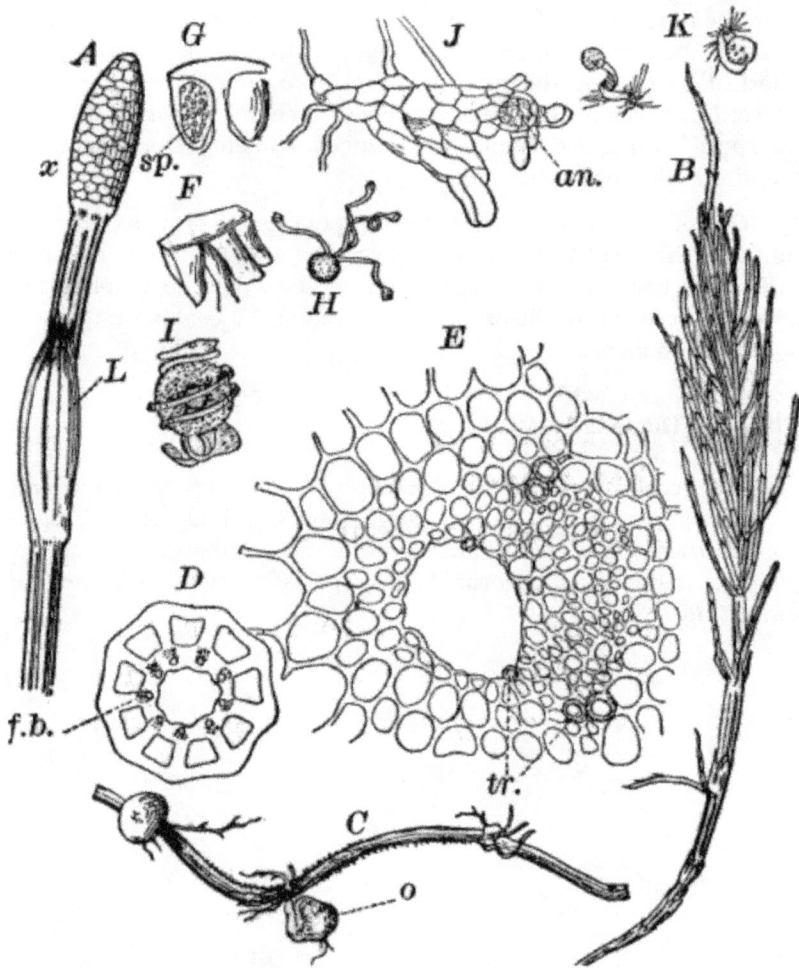

FIG. 72.—*A, spore-bearing stem of the field horse-tail (Equisetum), × 1. x, the spore-bearing cone. B, sterile stem of the same, × ½. C, underground stem, with tubers (o), × ½. D, cross-section of an aerial stem, × 5. f.b. fibro-vascular bundle. E, a single fibro-vascular bundle, × 150. tr. vessels. F, a single leaf from the cone, × 5. G, the same cut lengthwise, through a spore sac (sp.), × 5. H, a spore, × 50. I, the same, moistened so that the elaters are coiled up, × 150. J, a male prothallium, × 50. an. an antheridium. K, spermatozoids, × 300.*

One of the commonest forms is the field horse-tail (*Equisetum arvense*), a very abundant and widely distributed species. It grows in low, moist ground, and is often found in great abundance growing in the sand or gravel used as "ballast" for railway tracks.

The plant sends up branches of two kinds from a creeping underground stem that may reach a length of a metre or more. This stem (Fig. 72, C) is distinctly jointed, bearing at each joint a toothed sheath, best seen in the younger portions, as they are apt to be destroyed in the older parts. Sometimes attached to this are small tubers (o) which are much-shortened branches and under favorable circumstances give rise to new stems. They have a hard, brown rind, and are composed within mainly of a firm, white tissue, filled with starch.

The surface of the stem is marked with furrows, and a section across it shows that corresponding to these are as many large air spaces that traverse the stem from joint to joint. From the joints numerous roots, quite like those of the ferns, arise.

If the stem is dug up in the late fall or winter, numerous short branches of a lighter color will be found growing from the joints. These later grow up above ground into branches of two sorts. Those produced first (Fig. 72, A), in April or May, are stouter than the others, and nearly destitute of chlorophyll. They are usually twenty to thirty centimetres in height, of a light reddish brown color, and, like all the stems, distinctly jointed. The sheaths about the joints (L) are much larger than in the others, and have from ten to twelve large black teeth at the top. These sheaths are the leaves. At the top of the branch the joints are very close together, and the leaves of different form, and closely set so as to form a compact cone (x).

A cross-section of the stem (D) shows much the same structure as the underground stem, but the number of air spaces is larger, and in addition there is a large central cavity. The fibro-vascular bundles (f.b.) are arranged in a circle, alternating with the air channels, and each one has running through it a small air passage.

The cone at the top of the branch is made up of closely set, shield-shaped leaves, which are mostly six-sided, on account of the pressure. These leaves (F, G) have short stalks, and are arranged in circles about the stem. Each one has a number of spore cases hanging down from the edge, and opening by a cleft on the inner side (G, sp.). They are filled with a mass of greenish spores that shake out at the slightest jar when ripe.

The sterile branches (*B*) are more slender than the spore-bearing ones, and the sheaths shorter. Surrounding the joints, apparently just below the sheaths, but really breaking through their bases, are circles of slender branches resembling the main branch, but more slender. The sterile branches grow to a height of forty to fifty centimetres, and from their bushy form the popular name of the plant, "horse-tail," is taken. The surface of the plant is hard and rough, due to the presence of great quantities of flint in the epidermis,—a peculiarity common to all the species.

The stem is mainly composed of large, thin-walled cells, becoming smaller as they approach the epidermis. The outer cells of the ground tissue in the green branches contain chlorophyll, and the walls of some of them are thickened. The fibro-vascular bundles differ entirely from those of the ferns. Each bundle is nearly triangular in section (*E*), with the point inward, and the inner end occupied by a large air space. The tracheary tissue is only slightly developed, being represented by a few vessels [9] (*tr.*) at the outer angles of the bundle, and one or two smaller ones close to the air channel. The rest of the bundle is made up of nearly uniform, rather thin-walled, colorless cells, some of which, however, are larger, and have perforated cross-walls, representing the sieve tubes of the fern bundle. There is no individual bundle sheath, but the whole circle of bundles has a common outer sheath.

The epidermis is composed of elongated cells whose walls present a peculiar beaded appearance, due to the deposition of flint within them. The breathing pores are arranged in vertical lines, and resemble in general appearance those of the ferns, though differing in some minor details. Like the other epidermal cells the guard cells have heavy deposits of flint, which here are in the form of thick transverse bars.

The spore cases have thin walls whose cells, shortly before maturity, develop thickenings upon their walls, which have to do with the opening of the spore case. The spores (*H*, *I*) are round cells containing much chlorophyll and provided with four peculiar appendages called elaters. The elaters are extremely sensitive to changes in moisture, coiling up tightly when moistened (*I*), but quickly springing out again when dry (*H*). By dusting a few dry spores upon a slide, and putting it under the microscope without any water, the movement may be easily examined. Lightly breathing upon them will cause the

141

elaters to contract, but in a moment, as soon as the moisture of the breath has evaporated, they will uncoil with a quick jerk, causing the spores to move about considerably.

The fresh spores begin to germinate within about twenty-four hours, and the early stages, which closely resemble those of the ferns, may be easily followed by sowing the spores in water. With care it is possible to get the mature prothallia, which should be treated as described for the fern prothallia. Under favorable conditions, the first antheridia are ripe in about five weeks; the archegonia, which are borne on separate plants, a few weeks later. The antheridia (Fig. 72, *J, an.*) are larger than those of the ferns, and the spermatozoids (*K*) are thicker and with fewer coils, but otherwise much like fern spermatozoids.

The archegonia have a shorter neck than those of the ferns, and the neck is straight.

Both male and female prothallia are much branched and very irregular in shape.

There are a number of common species of *Equisetum*. Some of them, like the common scouring rush (*E. hiemale*), are unbranched, and the spores borne at the top of ordinary green branches; others have all the stems branching like the sterile stems of the field horsetail, but produce a spore-bearing cone at the top of some of them.

Class III. — The Club Mosses (*LYCOPODINÆ*).

The last class of the pteridophytes includes the ground pines, club mosses, etc., and among cultivated plants numerous species of the smaller club mosses (*Selaginella*).

Two orders are generally recognized, although there is some doubt as to the relationship of the members of the second order. The first order, the larger club mosses (*Lycopodiaceæ*) is represented in the northern states by a single genus (*Lycopodium*), of which the common ground pine (*L. dendroideum*) (Fig. 73) is a familiar species. The plant grows in the evergreen forests of the northern United States as well as in the mountains further south, and in the larger northern cities is often sold in large quantities at the holidays for decorating. It sends up from a creeping, woody, subterranean stem, numerous smaller stems which branch extensively, and are thickly set with small moss-like leaves, the whole looking much like a little tree. At the ends of some of the branches are small cones (*A, x, B*) composed of closely overlapping, scale-like leaves, much as in a fir cone. Near the base, on the inner surface of each of these scales, is a kidney-shaped capsule (*C, sp.*) opening by a cleft along the upper edge and filled with a mass of fine yellow powder. These capsules are the spore cases.

The bases of the upright stems are almost bare, but become covered with leaves higher up. The leaves are in shape like those of a moss, but are thicker. The spore-bearing leaves are broader and when slightly magnified show a toothed margin.

The stem is traversed by a central fibro-vascular cylinder that separates easily from the surrounding tissue, owing to the rupture of the cells of the bundle sheath, this being particularly frequent in dried specimens. When slightly magnified the arrangement of the tissues may be seen (Fig. 73, *E*). Within the epidermis is a mass of ground tissue of firm, woody texture surrounding the central oval or circular fibro-vascular cylinder. This shows a number of white bars (xylem) surrounded by a more delicate tissue (phloem).

On magnifying the section more strongly, the cells of the ground tissue (*G*) are seen to be oval in outline, with thick striated walls and small intercellular spaces. Examined in longitudinal sections they are long and pointed, belonging to the class of cells known as "fibres."

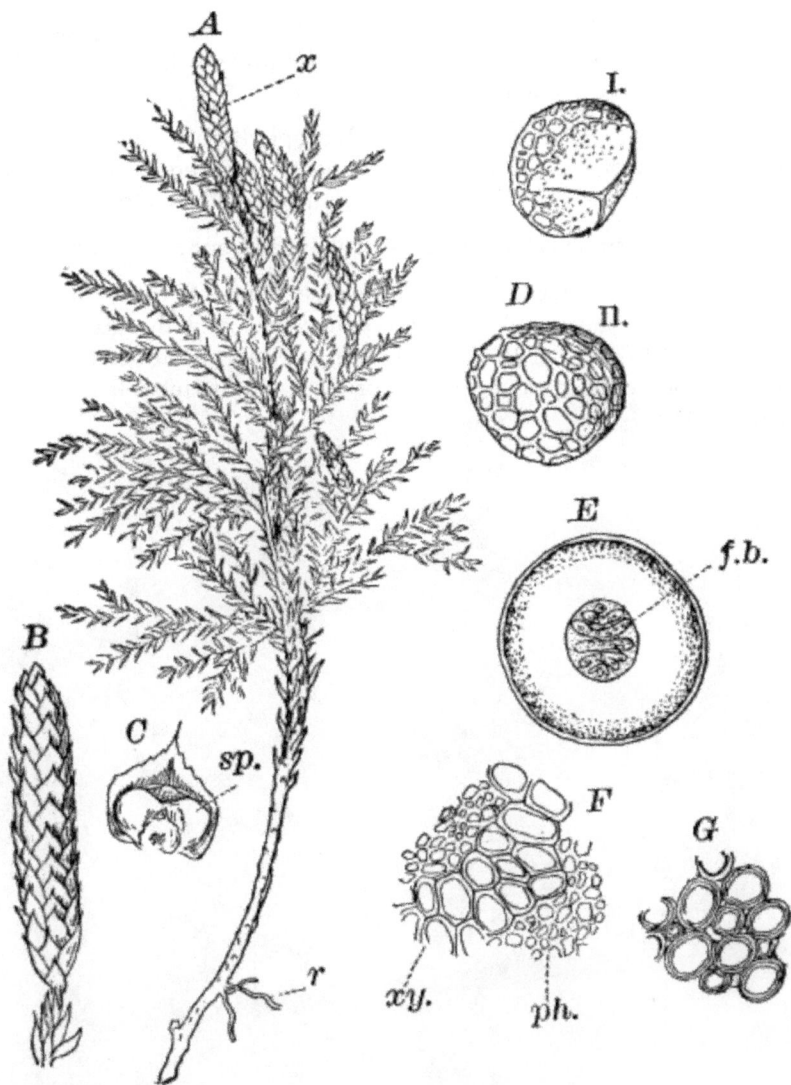

FIG. 73.—A, a club moss (Lycopodium), × ⅓. x, cone. r, root. B, a cone, × 1. C, single scale with sporangium (sp.). D, spores: I, from above; II, from below, × 325. E, cross section of stem, × 8. f.b. fibro-vascular bundle. F, portion of the fibro-vascular bundle, × 150. G, cells of the ground tissue, × 150.

The xylem (*F*, *xy.*) of the fibro-vascular bundle is composed of tracheids, much like those of the ferns; the phloem is composed of narrow cells, pretty much all alike.

The spores (*D*) are destitute of chlorophyll and have upon the outside a network of ridges, except on one side where three straight lines converge, the spore being slightly flattened between them.

Almost nothing is known of the prothallia of our native species.

The second order (*Ligulatæ*) is represented by two very distinct families: the smaller club mosses (*Selaginelleæ*) and the quill-worts (*Isoeteæ*). Of the former the majority are tropical, but are common in greenhouses where they are prized for their delicate moss-like foliage (Fig. 74, *A*).

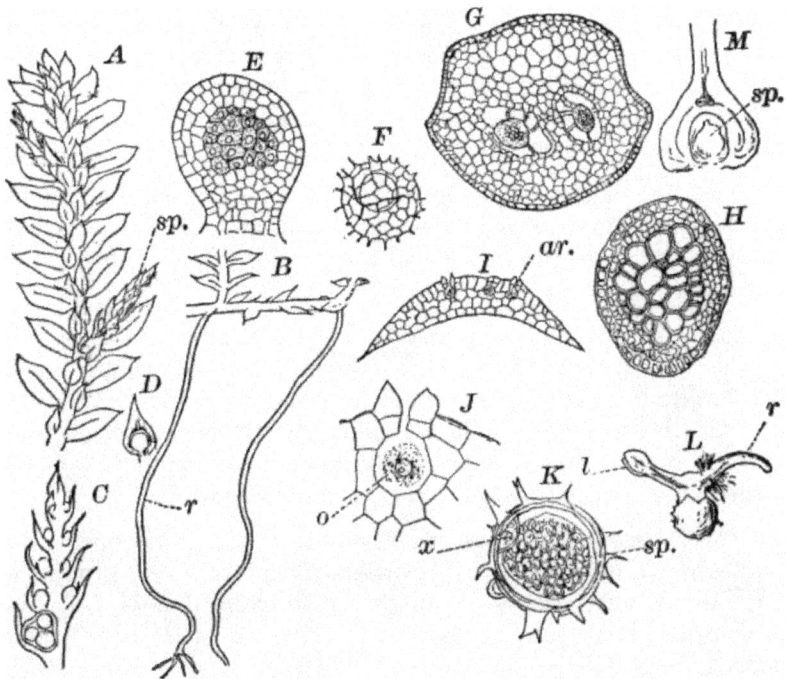

Fig. 74.—A, one of the smaller club mosses (Selaginella). sp. spore-bearing branch, × 2. B, part of a stem, sending down naked rooting branches (r), × 1. C, longitudinal section of a spike, with a single macrosporangium at the base; the others, microsporangia, × 3. D, a scale and microsporangium, × 5. E, young microsporangium, × 150. The shaded cells are the spore mother cells. F, a young macrospore, × 150. G, section of the stem, × 50. H, a single fibro-vascular bundle, × 150. I, vertical section of the female prothallium of Se-laginella, × 50. ar. archegonium. J, section of an open archegonium, × 300. o, the egg cell. K, microspore, with the contained male prothallium, × 300. x, vegetative cell. sp. sperm cells. L, young plant, with the attached macrospore, × 6. r, the first root. l, the first leaves.

The leaves in most species are like those of the larger club mosses, but more delicate. They are arranged in four rows on the up-per side of the stem, two being larger than the others. The smaller branches grow out sideways so that the whole branch appears flat-tened, reminding one of the habit of the higher liverworts. Special leafless branches (B, r) often grow downward from the lower side of the main branches, and on touching the ground develop roots which fork regularly.

146

The sporangia are much like those of the ground pines, and produced singly at the bases of scale leaves arranged in a spike or cone (A, sp.), but two kinds of spores, large and small, are formed. In the species figured the lower sporangium produces four large spores (macrospores); the others, numerous small spores (microspores).

Even before the spores are ripe the development of the prothallium begins, and this is significant, as it shows an undoubted relationship between these plants and the lowest of the seed plants, as we shall see when we study that group.

If ripe spores can be obtained by sowing them upon moist earth, the young plants will appear in about a month. The microspore (Fig. 74, K) produces a prothallium not unlike that of some of the water ferns, there being a single vegetative cell (x), and the rest of the prothallium forming a single antheridium. The spermatozoids are excessively small, and resemble those of the bryophytes.

The macrospore divides into two cells, a large lower one, and a smaller upper one. The latter gives rise to a flat disc of cells producing a number of small archegonia of simple structure (Fig. 74, I, J). The lower cell produces later a tissue that serves to nourish the young embryo.

The development of the embryo recalls in some particulars that of the seed plants, and this in connection with the peculiarities of the sporangia warrants us in regarding the *Ligulatæ* as the highest of existing pteridophytes, and to a certain extent connecting them with the lowest of the spermaphytes.

Resembling the smaller club mosses in their development, but differing in some important points, are the quill-worts (*Isoeteæ*). They are mostly aquatic forms, growing partially or completely submerged, and look like grasses or rushes. They vary from a few centimetres to half a metre in height. The stem is very short, and the long cylindrical leaves closely crowded together. The leaves which are narrow above are widely expanded and overlapping at the base. The spores are of two kinds, as in *Selaginella*, but the macrosporangia contain numerous macrospores. The very large sporangia (M, sp.) are in cavities at the bases of the leaves, and above each sporangium is a little pointed outgrowth (ligula), which is also found in the leaves of *Selaginella*. The quill-worts are not common plants, and owing to their habits of growth and resemblance to other plants, are likely to be overlooked unless careful search is made.

CHAPTER XIV.
SUB-KINGDOM VI.
Spermaphytes: Phænogams.

THE last and highest great division of the vegetable kingdom has been named *Spermaphyta*, "seed plants," from the fact that the structures known as seeds are peculiar to them. They are also commonly called flowering plants, though this name might be also appropriately given to certain of the higher pteridophytes.

In the seed plants the macrosporangia remain attached to the parent plant, in nearly all cases, until the archegonia are fertilized and the embryo plant formed. The outer walls of the sporangium now become hard, and the whole falls off as a seed.

In the higher spermaphytes the spore-bearing leaves (sporophylls) become much modified, and receive special names, those bearing the microspores being commonly known as stamens; those bearing the macrospores, carpels or carpophylls. The macrosporangia are also ordinarily known as "ovules," a name given before it was known that these were the same as the macrosporangia of the higher pteridophytes.

In addition to the spore-bearing leaves, those surrounding them may be much changed in form and brilliantly colored, forming, with the enclosed sporophylls, the "flower" of the higher spermaphytes.

As might be expected, the tissues of the higher spermaphytes are the most highly developed of all plants, though some of them are very simple. The plants vary extremely in size, the smallest being little floating plants, less than a millimetre in diameter, while others are gigantic trees, a hundred metres and more in height.

There are two classes of the spermaphytes: I., the Gymnosperms, or naked-seeded ones, in which the ovules (macrosporangia) are borne upon open carpophylls; and II., Angiosperms, covered-seeded plants, in which the carpophylls form a closed cavity (ovary) containing the ovules.

Class I. — Gymnosperms (*GYMNOSPERMÆ*).

The most familiar of these plants are the common evergreen trees (conifers), pines, spruces, cedars, etc. A careful study of one of these will give a good idea of the most important characteristics of the class, and one of the best for this purpose is the Scotch pine (*Pinus sylvestris*), which, though a native of Europe, is not infrequently met with in cultivation in America. If this species cannot be had by the student, other pines, or indeed almost any other conifer, will answer. The Scotch pine is a tree of moderate size, symmetrical in growth when young, with a central main shaft, and circles of branches at regular intervals; but as it grows older its growth becomes irregular, and the crown is divided into several main branches. [10] The trunk and branches are covered with a rough, scaly bark of a reddish brown color, where it is exposed by the scaling off of the outer layers. Covering the younger branches, but becoming thinner on the older ones, are numerous needle-shaped leaves. These are in pairs, and the base of each pair is surrounded by several dry, blackish scales. Each pair of leaves is really attached to a very short side branch, but this is so short as to make the leaves appear to grow directly from the main branch. Each leaf is about ten centimetres in length and two millimetres broad. Where the leaves are in contact they are flattened, but the outer side is rounded, so that a cross-section is nearly semicircular in outline. With a lens it is seen that there are five longitudinal lines upon the surface of the leaf, and careful examination shows rows of small dots corresponding to these. These dots are the breathing pores. If a cross-section is even slightly magnified it shows three distinct parts, — a whitish outer border, a bright green zone, and a central oval, colorless area, in which, with a little care, may be seen the sections of two fibro-vascular bundles. In the green zone are sometimes to be seen colorless spots, sections of resin ducts, containing the resin so characteristic of the tissues of the conifers.

The general structure of the stem may be understood by making a series of cross-sections through branches of different ages. In all, three regions are distinguishable; viz., an outer region (bark or cortex) (Fig. 76, *A*, *c*), composed in part of green cells, and, if the section has been made with a sharp knife, showing a circle of little openings, from each of which oozes a clear drop of resin. These are large resin ducts (*r*). The centre is occupied by a soft white tissue (pith), and the space between the pith and bark is filled by a mass of woody tissue. Traversing the wood are numerous radiating lines, some of which run from the bark to the pith, others only part way. These are called the medullary rays. While in sections from branches of any age these three regions are recognizable, their relative size varies extremely. In a section of a twig of the present year the bark and pith make up a considerable part of the section; but as older branches are examined, we find a rapid increase in the quantity of wood, while the thickness of the bark increases but slowly, and the pith scarcely at all. In the wood, too, each year's growth is marked by a distinct ring (A I, II). As the branches grow in diameter the outer bark becomes split and irregular, and portions die, becoming brown and hard.

The tree has a very perfect root system, but different from that of any pteridophytes. The first root of the embryo persists as the main or "tap" root of the full-grown tree, and from it branch off the secondary roots, which in turn give rise to others.

The sporangia are borne on special scale-like leaves, and arranged very much as in certain pteridophytes, notably the club mosses; but instead of large and small spores being produced near together, the two kinds are borne on special branches, or even on distinct trees (*e.g.* red cedar). In the Scotch pine the microspores are ripe about the end of May. The leaves bearing them are aggregated in small cones ("flowers"), crowded about the base of a growing shoot terminating the branches (Fig. 77, A ♂). The individual leaves (sporophylls) are nearly triangular in shape, and attached by the smaller end. On the lower side of each are borne two sporangia (pollen sacs) (C, *sp.*), opening by a longitudinal slit, and filled with innumerable yellow microspores (pollen spores), which fall out as a shower of yellow dust if the branch is shaken.

The macrosporangia (ovules) are borne on similar leaves, known as carpels, and, like the pollen sacs, borne in pairs, but on the upper side of the sporophyll instead of the lower. The female flowers appear when the pollen is ripe. The leaves of which they are composed are thicker than those of the male flowers, and of a pinkish color. At the base on the upper side are borne the two ovules (macrosporangia) (Fig. 77, E, *o*), and running through the centre is a ridge that ends in a little spine or point.

The ovule-bearing leaf has on the back a scale with fringed edge (F, *sc.*), quite conspicuous when the flower is young, but scarcely to be detected in the older cone. From the female flower is developed the cone (Fig. 75, A), but the process is a slow one, occupying two years. Shortly after the pollen is shed, the female flowers, which are at first upright, bend downward, and assume a brownish color, growing considerably in size for a short time, and then ceasing to grow for several months.

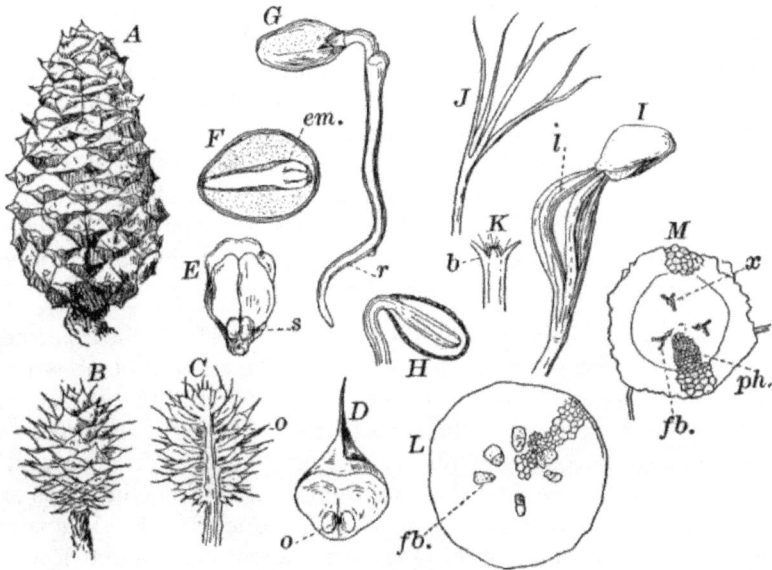

FIG. 75.—*Scotch pine (Pinus sylvestris). A, a ripe cone,* × ½. *B, a year-old cone,* × 1. *C, longitudinal section of B. D, a single scale of B, showing the sporangia (ovules) (o),* × 2. *E, a scale from a ripe cone, with the seeds (s),* × ½. *F, longitudinal section of a ripe seed,* × 3. *em. the embryo. G, a germinating seed,* × 2. *r, the primary root. H, longitudinal section through G, showing the first leaves of the young plant still surrounded by the endosperm,* × 4. *I, an older plant with the leaves (l) withdrawing from the seed coats,* × 4. *J, upper part of a young plant, showing the circle of primary leaves (cotyledons),* × 1. *K, section of the same,* × 2. *b, the terminal bud. L, cross-section of the stem of the young plant,* × 25. *fb. a fibro-vascular bundle. M, cross-section of the root,* × 25. *x, wood. ph. bast, of the fibro-vascular bundle.*

In Figure 75, *B*, is shown such a flower as it appears in the winter and early spring following. The leaves are thick and fleshy, closely pressed together, as is seen by dividing the flower lengthwise, and each leaf ends in a long point (*D*). The ovules are still very small. As the growth of the tree is resumed in the spring, the flower (cone) increases rapidly in size and becomes decidedly green in color, the ovules increasing also very much in size. If a scale from such a cone is examined about the first of June, the ovules will probably be nearly full-grown, oval, whitish bodies two to three millimetres in length. A careful longitudinal section of the scale through the ovule will show the general structure. Such a section is shown in Figure 77, *G*. Comparing this with the sporangia of the pteridophytes, the first difference that strikes us is the presence of an outer coat or integument (*in.*), which is absent in the latter. The single macrospore (*sp.*) is very large and does not lie free in the cavity of the sporangium, but is in close contact with its wall. It is filled with a colorless tissue, the prothallium, and if mature, with care it is possible to see, even with a hand lens, two or more denser oval bodies (*ar.*), the egg cells of the archegonia, which here are very large. The integument is not entirely closed at the top, but leaves a little opening through which the pollen spores entered when the flower was first formed.

After the archegonia are fertilized the outer parts of the ovule become hard and brown, and serve to protect the embryo plant, which reaches a considerable size before the sporangium falls off. As the walls of the ovule harden, the carpel or leaf bearing it undergoes a similar change, becoming extremely hard and woody, and as each one ends in a sharp spine, and they are tightly packed together, it is almost impossible to separate them. The ripe cone (Fig. 75, *A*) remains closed during the winter, but in the spring, about the time the flowers are mature, the scales open spontaneously and discharge the ripened ovules, now called seeds. Each seed (*E, s*) is surrounded by a membranous envelope derived from the scale to which it is attached, which becomes easily separated from the seed. The opening of the cones is caused by drying, and if a number of ripe cones are gathered in the winter or early spring, and allowed to dry in an ordinary room, they will in a day or two open, often with a sharp, crackling sound, and scatter the ripe seeds.

A section of a ripe seed (*F*) shows the embryo (*em.*) surrounded by a dense, white, starch-bearing tissue derived from the prothallium cells, and called the "endosperm." This fills up the whole seed which is surrounded by the hardened shell derived from the integument and wall of the ovule. The embryo is elongated with a circle of small leaves at the end away from the opening of the ovule toward which is directed the root of the embryo.

The seed may remain unchanged for months, or even years, without losing its vitality, but if the proper conditions are provided, the embryo will develop into a new plant. To follow the further growth of the embryo, the ripe seeds should be planted in good soil and kept moderately warm and moist. At the end of a week or two some of the seeds will probably have sprouted. The seed absorbs water, and the protoplasm of the embryo renews its activity, beginning to feed upon the nourishing substances in the cells of the endosperm. The embryo rapidly increases in length, and the root pushes out of the seed growing rapidly downward and fastening itself in the soil (G, r). Cutting the seed lengthwise we find that the leaves have increased much in length and become green (one of the few cases where chlorophyll is formed in the absence of light). As these leaves (called "cotyledons" or seed leaves) increase in length, they gradually withdraw from the seed whose contents they have exhausted, and the young plant enters upon an independent existence.

The young plant has a circle of leaves, about six in number, surrounding a bud which is the growing point of the stem, and in many conifers persists as long as the stem grows (Fig. 75, K, b). A cross-section of the young stem shows about six separate fibro-vascular bundles arranged in a circle (S, fb.). The root shows a central fibro-vascular cylinder surrounded by a dark-colored ground tissue. Growing from its surface are numerous root hairs (Fig. 75, M).

For examining the microscopic structure of the pine, fresh material is for most purposes to be preferred, but alcoholic material will answer, and as the alcohol hardens the resin, it is for that reason preferable.

Cross-sections of the leaf, when sufficiently magnified, show that the outer colorless border of the section is composed of two parts: the epidermis of a single row of regular cells with very thick outer walls, and irregular groups of cells lying below them. These latter have thick walls appearing silvery and clearer than the epidermal cells. They vary a good deal, in some leaves being reduced to a single row, in others forming very conspicuous groups of some size. The green tissue of the leaf is much more compact than in the fern we examined, and the cells are more nearly round and the intercellular spaces smaller. The chloroplasts are numerous and nearly round in shape.

Scattered through the green tissue are several resin passages (r), each surrounded by a circle of colorless, thick-walled cells, like those under the epidermis. At intervals in the latter are openings— breathing pores—(Fig. 76, J), below each of which is an intercellular space (*i*). They are in structure like those of the ferns, but the walls of the guard cells are much thickened like the other epidermal cells.

Each leaf is traversed by two fibro-vascular bundles of entirely different structure from those of the ferns. Each is divided into two nearly equal parts, the wood (*x*) lying toward the inner, flat side of the leaf, the bast (*T*) toward the outer, convex side. This type of bundle, called "collateral," is the common form found in the stems and leaves of seed plants. The cells of the wood or xylem are rather larger than those of the bast or phloem, and have thicker walls than any of the phloem cells, except the outermost ones which are thick-walled fibres like those under the epidermis. Lying between the bundles are comparatively large colorless cells, and surrounding the whole central area is a single line of cells that separates it sharply from the surrounding green tissue.

In longitudinal sections, the cells, except of the mesophyll (green tissue) are much elongated. The mesophyll cells, however, are short and the intercellular spaces much more evident than in the cross-section. The colorless cells have frequently rounded depressions or pits upon their walls, and in the fibro-vascular bundle the difference between the two portions becomes more obvious. The wood is distinguished by the presence of vessels with close, spiral or ring-shaped thickenings, while in the phloem are found sieve tubes, not unlike those in the ferns.

The fibro-vascular bundles of the stem of the seedling plant show a structure quite similar to that of the leaf, but very soon a difference is manifested. Between the two parts of the bundle the cells continue to divide and add constantly to the size of the bundle, and at the same time the bundles become connected by a line of similar growing cells, so that very early we find a ring of growing cells extending completely around the stem. As the cells in this ring increase in number, owing to their rapid division, those on the borders of the ring lose the power of dividing, and gradually assume the character of the cells on which they border (Fig. 76, *B*, *cam*.). The growth on the inside of the ring is more rapid than on the outer border, and the ring continues comparatively near the surface of the stem (Fig. 76, *A*, *cam*.). The spaces between the bundles do not increase materially in breadth, and as the bundles increase in size become in comparison very small, appearing in older stems as mere lines between the solid masses of wood that make up the inner portion of the bundles. These are the primary medullary rays, and connect the pith in the centre of the stem with the bark. Later, similar plates of cells are formed, often only a single cell thick, and appearing when seen in cross-section as a single row of elongated cells (*C, m*).

As the stem increases in diameter the bundles become broader and broader toward the outside, and taper to a point toward the centre, appearing wedge-shaped, the inner ends projecting into the pith. The outer limits of the bundles are not nearly so distinct, and it is not easy to tell when the phloem of the bundles ends and the ground tissue of the bark begins.

A careful examination of a cross-section of the bark shows first, if taken from a branch not more than two or three years old, the epidermis composed of cells not unlike those of the leaf, but whose walls are usually browner. Underneath are cells with brownish walls, and often more or less dry and dead. These cells give the brown color to the bark, and later both epidermis and outer ground tissue become entirely dead and disappear. The bulk of the ground tissue is made up of rather large, loose cells, the outer ones containing a good deal of chlorophyll. Here and there are large resin ducts (Fig. 76, *H*), appearing in cross-section as oval openings surrounded by several concentric rows of cells, the innermost smaller and with denser contents. These secrete the resin that fills the duct and oozes out when the stem is cut. All of the cells of the bark contain more or less starch.

The phloem, when strongly magnified, is seen to be made up of cells arranged in nearly regular radiating rows. Their walls are not very thick and the cells are usually somewhat flattened in a radial direction.

Some of the cells are larger than the others, and these are found to be, when examined in longitudinal section, sieve tubes (Fig. 76, *E*) with numerous lateral sieve plates quite similar to those found in the stems of ferns.

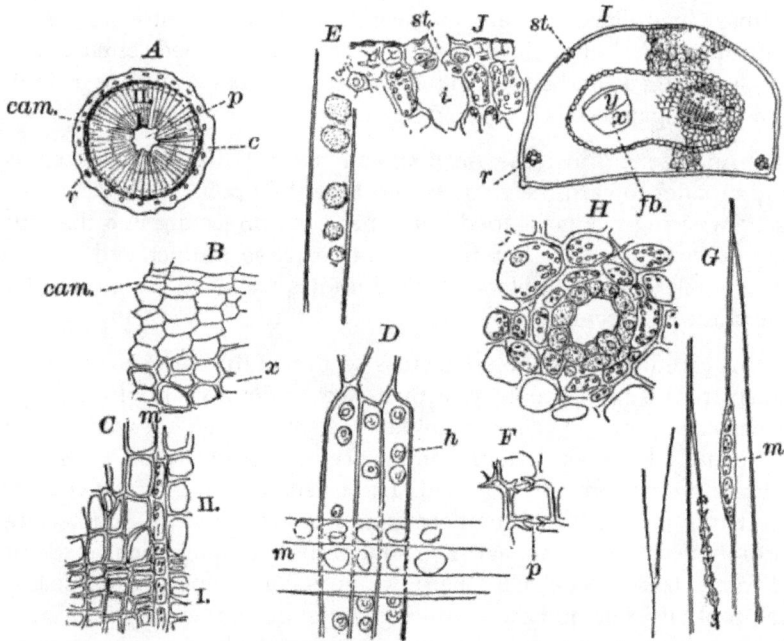

FIG. 76.—*Scotch pine. A, cross-section of a two-year-old branch, × 3. p, pith. c, bark. The radiating lines are medullary rays. r, resin ducts. B, part of the same, × 150. cam. cambium cells. x, tracheids. C, cross-section of a two-year-old branch at the point where the two growth rings join: I, the cells of the first year's growth; II, those of the second year. m, a medullary ray, × 150. D, longitudinal section of a branch, showing the form of the tracheids and the bordered pits upon their walls. m, medullary ray, × 150. E, part of a sieve tube, × 300. F, cross-section of a tracheid passing through two of the pits in the wall (p), × 300. G, longitudinal section of a branch, at right angles to the medullary rays (m). At y, the section has passed through the wall of a tracheid, bearing a row of pits, × 150. H, cross-section of a resin duct, × 150. I, cross-section of a leaf, × 20. fb. fibro-vascular bundle. r, resin duct. J, section of a breathing pore, × 150. i, the air space below it.*

The growing tissue (cambium), separating the phloem from the wood, is made up of cells quite like those of the phloem, into which they insensibly merge, except that their walls are much thinner, as is always the case with rapidly growing cells. These cells (B, cam.) are arranged in radial rows and divide, mainly by walls, at right angles to the radii of the stem. If we examine the inner side of the ring, the change the cells undergo is more marked. They become of nearly equal diameter in all directions, and the walls become woody, showing at the same time distinct stratification (*B*, *x*).

On examining the xylem, where two growth rings are in contact, the reason of the sharply marked line seen when the stem is examined with the naked eye is obvious. On the inner side of this line (*I*), the wood cells are comparatively small and much flattened, while the walls are quite as heavy as those of the much larger cells (*II*) lying on the outer side of the line. The small cells show the point where growth ceased at the end of the season, the cells becoming smaller as growth was feebler. The following year when growth commenced again, the first wood cells formed by the cambium were much larger, as growth is most vigorous at this time, and the wood formed of these larger cells is softer and lighter colored than that formed of the smaller cells of the autumn growth.

The wood is mainly composed of tracheids, there being no vessels formed except the first year. These tracheids are characterized by the presence of peculiar pits upon their walls, best seen when thin longitudinal sections are made in a radial direction. These pits (Fig. 76, *D*, *p*) appear in this view as double circles, but if cut across, as often happens in a cross-section of the stem, or in a longitudinal section at right angles to the radius (tangential), they are seen to be in shape something like an inverted saucer with a hole through the bottom. They are formed in pairs, one on each side of the wall of adjacent tracheids, and are separated by a very delicate membrane (*F*, *p*, *G*, *y*). These "bordered" pits are very characteristic of the wood of all conifers.

The structure of the root is best studied in the seedling plant, or in a rootlet of an older one. The general plan of the root is much like that of the pteridophytes. The fibro-vascular bundle (Fig. 75, *M*, *fb.*) is of the so-called radial type, there being three xylem masses (*x*) alternating with as many phloem masses (*ph.*) in the root of the seedling. This regularity becomes destroyed as the root grows older by the formation of a cambium ring, something like that in the stem.

The development of the sporangia is on the whole much like that of the club mosses, and will not be examined here in detail. The microspores (pollen spores) are formed in groups of four in precisely the same way as the spores of the bryophytes and pteridophytes, and by collecting the male flowers as they begin to appear in the spring, and crushing the sporangia in water, the process of division may be seen. For more careful examination they may be crushed in a mixture of water and acetic acid, to which is added a little gentian violet. This mixture fixes and stains the nuclei of the spores, and very instructive preparations may thus be made. [11]

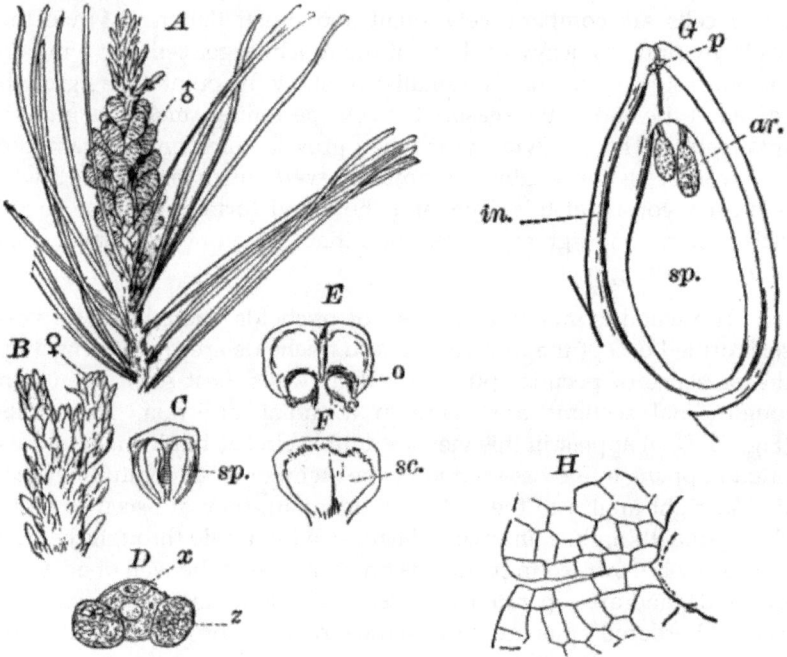

FIG. 77. — *Scotch pine (except E and F). A, end of a branch bearing a cluster of male flowers (♂), × ½. B, a similar branch, with two young female flowers (♀), natural size. C, a scale from a male flower, showing the two sporangia (sp.); × 5. D, a single ripe pollen spore (microspore), showing the vegetative cell (x), × 150. E, a similar scale, from a female flower of the Austrian pine, seen from within, × 4. o, the sporangium (ovule). F, the same, seen from the back, showing the scale (sc.) attached to the back. G, longitudinal section through a full-grown ovule of the Scotch pine. p, a pollen spore sending down its tube to the archegonia (ar.). sp. the prothallium (endosperm), filling up the embryo sac, × 10. H, the neck of the archegonium, × 150.*

The ripe pollen spores (Fig. 77, *D*) are oval cells provided with a double wall, the outer one giving rise to two peculiar bladder-like appendages (*z*). Like the microspores of the smaller club mosses, a

159

small cell is cut off from the body of the spore (*x*). These pollen spores are carried by the wind to the ovules, where they germinate.

The wall of the ripe sporangium or pollen sac is composed of a single layer of cells in most places, and these cells are provided with thickened ridges which have to do with opening the pollen sac.

We have already examined in some detail the structure of the macrosporangium or ovule. In the full-grown ovule the macrospore, which in the seed plants is generally known as the "embryo sac," is completely filled with the prothallium or "endosperm." In the upper part of the prothallium several large archegonia are formed in much the same way as in the pteridophytes. The egg cell is very large, and appears of a yellowish color, and filled with large drops that give it a peculiar aspect. There is a large nucleus, but it is not always readily distinguished from the other contents of the egg cell. The neck of the archegonium is quite long, but does not project above the surface of the prothallium (Fig. 77, *H*).

The pollen spores are produced in great numbers, and many of them fall upon the female flowers, which when ready for pollination have the scales somewhat separated. The pollen spores now sift down to the base of the scales, and finally reach the opening of the ovule, where they germinate. No spermatozoids are produced, the seed plants differing in this respect from all pteridophytes. The pollen spore bursts its outer coat, and sends out a tube which penetrates for some distance into the tissue of the ovule, acting very much as a parasitic fungus would do, and growing at the expense of the tissue through which it grows. After a time growth ceases, and is not resumed until the development of the female prothallium and archegonia is nearly complete, which does not occur until more than a year from the time the pollen spore first reaches the ovule. Finally the pollen tube penetrates down to and through the open neck of the archegonium, until it comes in contact with the egg cell. These stages can only be seen by careful sections through a number of ripe ovules, but the track of the pollen tube is usually easy to follow, as the cells along it are often brown and apparently dead (Fig. 77, *G*).

Classification of the Gymnosperms.

There are three classes of the gymnosperms: I., cycads (*Cycadeæ*); II., conifers (*Coniferæ*); III., joint firs (*Gnetaceæ*). All of the gymnosperms of the northern United States belong to the second order, but representatives of the others are found in the southern and southwestern states.

The cycads are palm-like forms having a single trunk crowned by a circle of compound leaves. Several species are grown for ornament in conservatories, and a few species occur native in Florida, but otherwise do not occur within our limits.

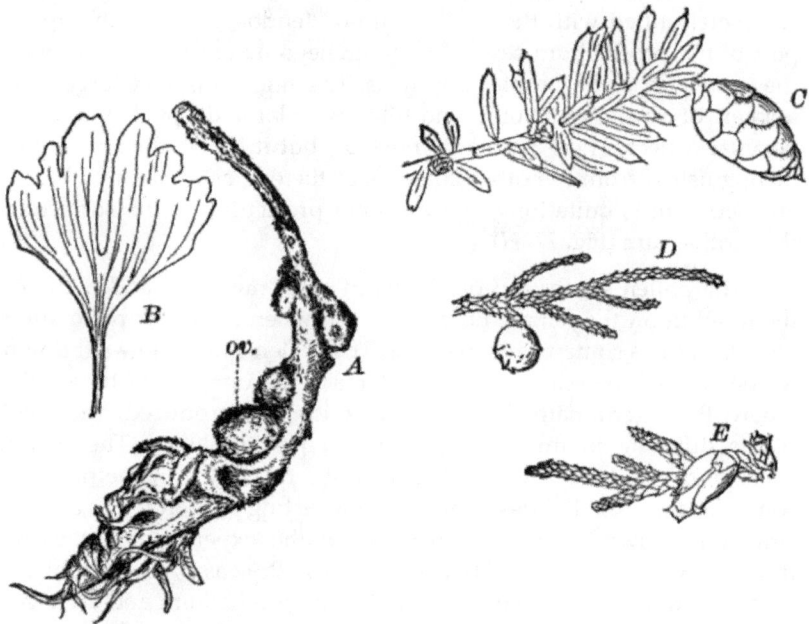

FIG. 78.—*Illustrations of gymnosperms. A, fruiting leaf of a cycad (Cycas), with macrosporangia (ovules) (ov.), × ¼. B, leaf of Gingko, × ½. C, branch of hemlock (Tsuga), with a ripe cone, × 1. D, red cedar (Juniperus), × 1. E, Arborvitæ (Thuja), × 1.*

The spore-bearing leaves usually form cones, recalling somewhat in structure those of the horse-tails, but one of the commonest cultivated species (*Cycas revoluta*) bears the ovules, which are very large, upon leaves that are in shape much like the ordinary ones (Fig. 78, A).

Of the conifers, there are numerous familiar forms, including all our common evergreen trees. There are two sub-orders,—the true conifers and the yews. In the latter there is no true cone, but the

161

ovules are borne singly at the end of a branch, and the seed in the yew (*Taxus*) is surrounded by a bright red, fleshy integument. One species of yew, a low, straggling shrub, occurs sparingly in the northern states, and is the only representative of the group at the north. The European yew and the curious Japanese *Gingko* (Fig. 78, *B*) are sometimes met with in cultivation.

Of the true conifers, there are a number of families, based on peculiarities in the leaves and cones. Some have needle-shaped leaves and dry cones like the firs, spruces, hemlock (Fig. 78, *C*). Others have flattened, scale-like leaves, and more or less fleshy cones, like the red cedar (Fig. 78, *D*) and *Arbor-vitæ* (*E*).

A few of the conifers, such as the tamarack or larch (*Larix*) and cypress (*Taxodium*), lose their leaves in the autumn, and are not, therefore, properly "evergreen."

The conifers include some of the most valuable as well as the largest of trees. Their timber, especially that of some of the pines, is particularly valuable, and the resin of some of them is also of much commercial importance. Here belong the giant red-woods (*Sequoia*) of California, the largest of all American trees.

The joint firs are comparatively small plants, rarely if ever reaching the dimensions of trees. They are found in various parts of the world, but are few in number, and not at all likely to be met with by the ordinary student. Their flowers are rather more highly differentiated than those of the other gymnosperms, and are said to show some approach in structure to those of the angiosperms.

CHAPTER XV.
SPERMAPHYTES.

Class II. — Angiosperms.

The angiosperms include an enormous assemblage of plants, all those ordinarily called "flowering plants" belonging here. There is almost infinite variety shown in the form and structure of the tissues and organs, this being particularly the case with the flowers. As already stated, the ovules, instead of being borne on open carpels, are enclosed in a cavity formed by a single closed carpel or several united carpels. To the organ so formed the name "pistil" is usually applied, and this is known as "simple" or "compound," as it is composed of one or of two or more carpels. The leaves bearing the pollen spores are also much modified, and form the so-called "stamens." In addition to the spore-bearing leaves there are usually other modified leaves surrounding them, these being often brilliantly colored and rendering the flower very conspicuous. To these leaves surrounding the sporophylls, the general name of "perianth" or "perigone" is given. The perigone has a twofold purpose, serving both to protect the sporophylls, and, at least in bright-colored flowers, to attract insects which, as we shall see, are important agents in transferring pollen from one flower to another.

When we compare the embryo sac (macrospore) of the angiosperms with that of the gymnosperms a great difference is noticed, there being much more difference than between the latter and the higher pteridophytes. Unfortunately there are very few plants where the structure of the embryo sac can be readily seen without very skilful manipulation.

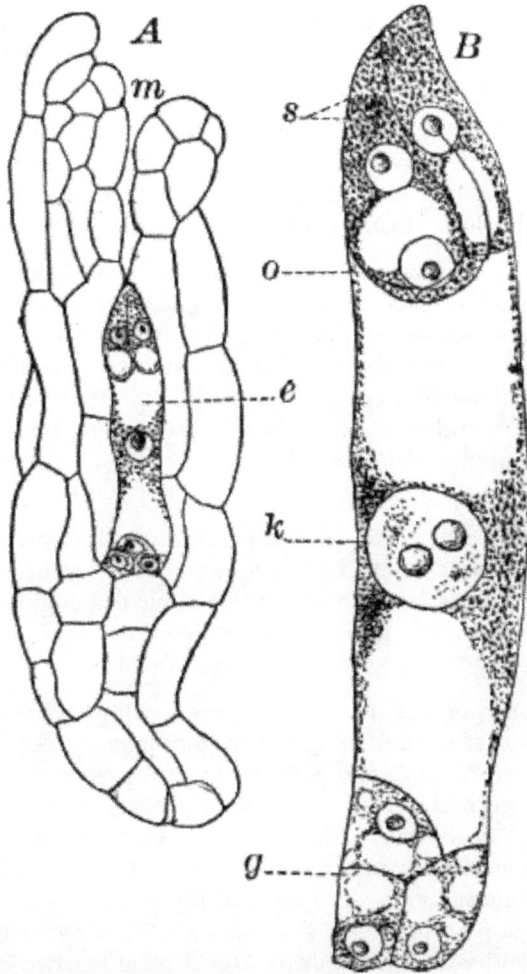

FIG. 79.—*A, ripe ovule of Monotropa uniflora, in optical section, × 100. m, micropyle. e, embryo sac. B, the embryo sac, × 300. At the top is the egg apparatus, consisting of the two synergidæ (s), and the egg cell (o). In the centre is the "endosperm nucleus" (k). At the bottom, the "antipodal cells" (g).*

There are, however, a few plants in which the ovules are very small and transparent, so that they may be mounted whole and examined alive. The best plant for this purpose is probably the "Indian pipe" or "ghost flower," a curious plant growing in rich woods, blossoming in late summer. It is a parasite or saprophyte, and entirely destitute of chlorophyll, being pure white throughout. It bears a single nodding flower at the summit of the stem. (Another species much like it, but having several brownish flowers, is shown in Figure 115, L.)

If this plant can be had, the structure of the ovule and embryo sac may be easily studied, by simply stripping away the tissue bearing the numerous minute ovules, and mounting a few of them in water, or water to which a little sugar has been added.

The ovules are attached to a stalk, and each consists of about two layers of colorless cells enclosing a central, large, oblong cell (Fig. 79, A, E), the embryo sac or macrospore. If the ovule is from a flower that has been open for some time, we shall find in the centre of the embryo sac a large nucleus (k) (or possibly two which afterward unite into one), and at each end three cells. Those at the base (g) probably represent the prothallium, and those at the upper end a very rudimentary archegonium, here generally called the "egg apparatus."

Of the three cells of the "egg apparatus" the lower (o) one is the egg cell; the others are called "synergidæ." The structure of the embryo sac and ovules is quite constant among the angiosperms, the differences being mainly in the shape of the ovules, and the degree to which its coverings or integuments are developed.

The pollen spores of many angiosperms will germinate very easily in a solution of common sugar in water: about fifteen per cent of sugar is the best. A very good plant for this purpose is the sweet pea, whose pollen germinates very rapidly, especially in warm weather. The spores may be sown in a little of the sugar solution in any convenient vessel, or in a hanging drop suspended in a moist chamber, as described for germinating the spores of the slime moulds. The tube begins to develop within a few minutes after the spores are placed in the solution, and within an hour or so will have reached a considerable length. Each spore has two nuclei, but they are less evident here than in some other forms (Fig. 79).

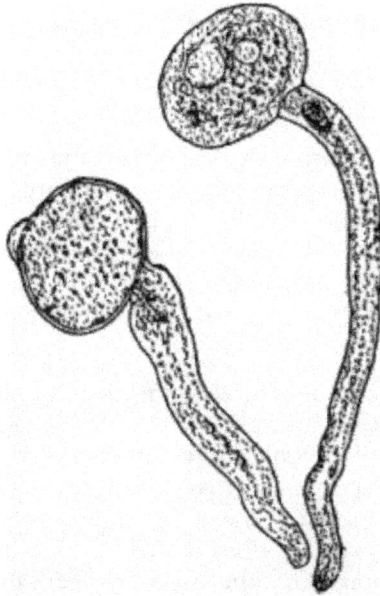

FIG. 80. — Germinating pollen spores of the sweet pea, × 200.

The upper part of the pistil is variously modified, having either little papillæ which hold the pollen spores, or are viscid. In either case the spores germinate when placed upon this receptive part (stigma) of the pistil, and send their tubes down through the tissues of the pistil until they reach the ovules, which are fertilized much as in the gymnosperms.

The effect of fertilization extends beyond the ovule, the ovary and often other parts of the flower being affected, enlarging and often becoming bright-colored and juicy, forming the various fruits of the angiosperms. These fruits when ripe may be either dry, as in the case of grains of various kinds, beans, peas, etc.; or the ripe fruit may be juicy, serving in this way to attract animals of many kinds which feed on the juicy pulp, and leave the hard seeds uninjured, thus helping to distribute them. Common examples of these fleshy fruits are offered by the berries of many plants; apples, melons, cherries, etc., are also familiar examples.

The seeds differ a good deal both in regard to size and the degree to which the embryo is developed at the time the seed ripens.

Classification of the Angiosperms.

The angiosperms are divided into two sub-classes: I. *Monocotyledons* and II. *Dicotyledons.*

The monocotyledons comprise many familiar plants, both ornamental and useful. They have for the most part elongated, smooth-edged leaves with parallel veins, and the parts of the flower are in threes in the majority of them. As their name indicates, there is but one cotyledon or seed leaf, and the leaves from the first are alternate. As a rule the embryo is very small and surrounded by abundant endosperm.

The most thoroughly typical members of the sub-class are the lilies and their relatives. The one selected for special study here, the yellow adder-tongue, is very common in the spring; but if not accessible, almost any liliaceous plant will answer. Of garden flowers, the tulip, hyacinth, narcissus, or one of the common lilies may be used; of wild flowers, the various species of *Trillium* (Fig. 83, *A*) are common and easily studied forms, but the leaves are not of the type common to most monocotyledons.

The yellow adder-tongue (*Erythronium americanum*) (Fig. 81) is one of the commonest and widespread of wild flowers, blossoming in the northern states from about the middle of April till the middle of May. Most of the plants found will not be in flower, and these send up but a single, oblong, pointed leaf. The flowering plant has two similar leaves, one of which is usually larger than the other. They seem to come directly from the ground, but closer examination shows that they are attached to a stem of considerable length entirely buried in the ground. This arises from a small bulb (*B*) to whose base numerous roots (*r*) are attached. Rising from between the leaves is a slender, leafless stalk bearing a single, nodding flower at the top.

The leaves are perfectly smooth, dull purplish red on the lower side, and pale green with purplish blotches above. The epidermis may be very easily removed, and is perfectly colorless. Examined closely, longitudinal rows of whitish spots may be detected: these are the breathing pores.

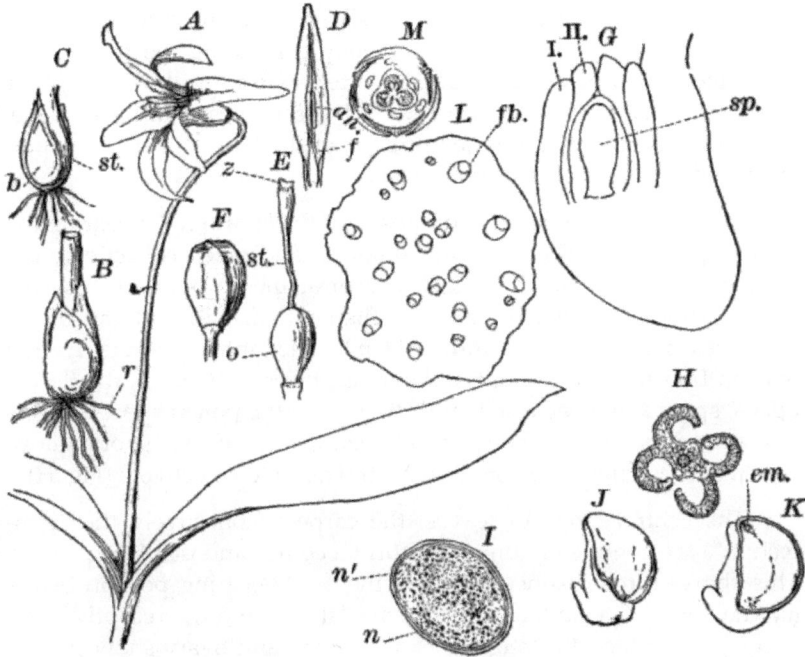

FIG. 81.—A, plant of the yellow adder-tongue (Erythronium americanum), × ⅓. B, the bulb of the same, × ½. r, roots. C, section of B. st. the base of the stem bearing the bulb for next year (b) at its base. D, a single petal and stamen, × ½. f, the filament. an. anther. E, the gynœcium (pistil), × 1. o, ovary. st. style. z, stigma. F, a full-grown fruit, × ½. G, section of a full-grown macrosporangium (ovule), × 25: I, II, the two integuments. sp. macrospore (embryo sac). H, cross-section of the ripe anther, × 12. I, a single pollen spore, × 150, showing the two nuclei (n, n□). J, a ripe seed, × 2. K, the same, in longitudinal section. em. the embryo. L, cross-section of the stem, × 12. fb. fibro-vascular bundle. M, diagram of the flower.

A cross-section of the stem shows numerous whitish areas scattered through it. These are the fibro-vascular bundles which in the monocotyledons are of a simple type. The bulb is composed of thick scales, which are modified leaves, and on cutting it lengthwise, we shall probably find the young bulb of next year (Fig. C, b) already forming inside it, the young bulb arising as a bud at the base of the stem of the present year.

The flower is made up of five circles of very much modified leaves, three leaves in each set. The two outer circles are much alike,

168

but the three outermost leaves are slightly narrower and strongly tinged with red on the back, completely concealing the three inner ones before the flower expands. The latter are pure yellow, except for a ridge along the back, and a few red specks near the base inside. These six leaves constitute the perigone of the flower; the three outer are called sepals, the inner ones petals.

The next two circles are composed of the sporophylls bearing the pollen spores. [12] These are the stamens, and taken collectively are known as the "Andrœcium." Each leaf or stamen consists of two distinct portions, a delicate stalk or "filament" (D, f), and the upper spore-bearing part, the "anther" (an.). The anther in the freshly opened flower has a smooth, red surface; but shortly after, the flower opens, splits along each side, and discharges the pollen spores. A section across the anther shows it to be composed of four sporangia or pollen sacs attached to a common central axis ("connective") (Fig. H).

The central circle of leaves, the carpels (collectively the "gynœcium") are completely united to form a compound pistil (Fig. 81, E). This shows three distinct portions, the ovule-bearing portion below (o), the "ovary," a stalk above (st.), the "style," and the receptive portion (z) at the top, the "stigma." Both stigma and ovary show plainly their compound nature, the former being divided into three lobes, the latter completely divided into three chambers, as well as being flattened at the sides with a more or less decided seam at the three angles. The ovules, which are quite large, are arranged in two rows in each chamber of the ovary, attached to the central column ("placenta").

The flowers open for several days in succession, but only when the sun is shining. They are visited by numerous insects which carry the pollen from one flower to another and deposit it upon the stigma, where it germinates, and the tube, growing down through the long style, finally reaches the ovules and fertilizes them. Usually only a comparatively small number of the seeds mature, there being almost always a number of imperfect ones in each pod. The pod or fruit (F) is full-grown about a month after the flower opens, and finally separates into three parts, and discharges the seeds. These are quite large (Fig. 81, J) and covered with a yellowish brown outer coat, and provided with a peculiar, whitish, spongy appendage attaching it to the placenta. A longitudinal section of a ripe seed (K) shows the very small, nearly triangular embryo (em.), while the rest of the cavity of the seed is filled with a white, starch-bearing tissue, the endosperm.

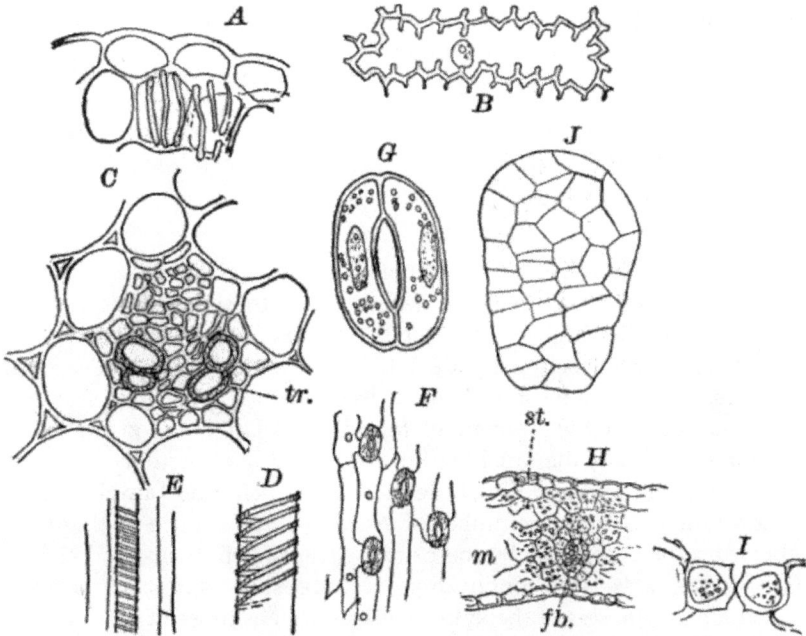

FIG. 82.—*Erythronium. A, a portion of the wall of the anther, × 150. B, a single epidermal cell from the petal, × 150. C, cross-section of a fibro-vascular bundle of the stem, × 150. tr. vessels. D, E, longitudinal section of the same, showing the markings of the vessels, × 150. F, a bit of the epidermis from the lower surface of a leaf, showing the breathing pores, × 50. G, a single breathing pore, × 200. H, cross-section of a leaf, × 50. st. a breathing pore. m, the mesophyll. fb. a vein. I, cross-section of a breathing pore, × 200. J, young embryo, × 150.*

A microscopical examination of the tissues of the plant shows them to be comparatively simple, this being especially the case with the fibro-vascular system.

The epidermis of the leaf is readily removed, and examination shows it to be made up of oblong cells with large breathing pores in rows. The breathing pores are much larger than any we have yet seen, and are of the type common to most angiosperms. The ordinary epidermal cells are quite destitute of chlorophyll, but the two cells (guard cells) enclosing the breathing pore contain numerous chloroplasts, and the oblong nuclei of these cells are usually conspicuous (Fig. 82, G). By placing a piece of the leaf between pieces of pith, and making a number of thin cross-sections at right angles to the longer axis of the

170

leaf, some of the breathing pores will probably be cut across, and their structure may be then better understood. Such a section is shown in Figure 82, *I*.

The body of the leaf is made up of chlorophyll-bearing cells of irregular shape and with large air spaces between (*H*, *m*). The veins traversing this tissue are fibro-vascular bundles of a type structure similar to that of the stem, which will be described presently.

The stem is made up principally of large cells with thin walls, which in cross-section show numerous small, triangular, intercellular spaces (*i*) at the angles. These cells contain, usually, more or less starch. The fibro-vascular bundles (*C*) are nearly triangular in section, and resemble considerably those of the field horse-tail, but they are not penetrated by the air channel, found in the latter. The xylem, as in the pine, is toward the outside of the stem, but the boundary between xylem and phloem is not well defined, there being no cambium present. In the xylem are a number of vessels (*C*, *tr*.) at once distinguishable from the other cells by their definite form, firm walls, and empty cavity. The vessels in longitudinal sections show spiral and ringed thickenings. The rest of the xylem cells, as well as those of the phloem, are not noticeably different from the cells of the ground tissue, except for their much smaller size, and absence of intercellular spaces.

The structure of the leaves of the perigone is much like that of the green leaves, but the tissues are somewhat reduced. The epidermis of the outer side of the sepals has breathing pores, but these are absent from their inner surface, and from both sides of the petals. The walls of the epidermal cells of the petals are peculiarly thickened by apparent infoldings of the wall (*B*), and these cells, as well as those below them, contain small, yellow bodies (chromoplasts) to which the bright color of the flower is due. The red specks on the base of the perigone leaves, as well as the red color of the back of the sepals, the stalk, and leaves are due to a purplish red cell sap filling the cells at these points.

The filaments or stalks of the stamens are made up of very delicate colorless cells, and the centre is traversed by a single fibro-vascular bundle, which is continued up through the centre of the anther. To study the latter, thin cross-sections should be made and mounted in water. Each of the four sporangia, or pollen sacs, is surrounded on the outside by a wall, consisting of two layers of cells, becoming thicker in the middle of the section where the single fibro-vascular bundle is seen (Fig. 81, *H*). On opening, the cavities of the adjacent sporangia are thrown together. The inner cells of the wall are marked by thickened bars, much as we saw in the pine (Fig. 82, *A*), and which, like these, are formed shortly before the pollen sacs open. The pollen spores (Fig. 81, *I*) are large, oval cells, having a double wall, the outer one somewhat heavier than the inner one, but sufficiently transparent to allow a clear view of the interior, which is filled with very dense, granular protoplasm in which may be dimly seen two nuclei (*n, ni.*), showing that here also there is a division of the spore contents, although no wall is present. The spores do not germinate very readily, and are less favorable for this purpose than those of some other monocotyledons. Among the best for this purpose are the spiderwort (*Tradescantia*) and *Scilla*.

Owing to the large size and consequent opacity of the ovules, as well as to the difficulty of getting the early stages, the development and finer structure of the ovule will not be discussed here. The full-grown ovule may be readily sectioned, and a general idea of its structure obtained. A little potash may be used to advantage in this study, carefully washing it away when the section is sufficiently cleared. We find now that the ovule is attached to a stalk (funiculus) (Fig. 81, *G, f*), the body of the ovule being bent up so as to lie against the stalk. Such an inverted ovule is called technically, "anatropous." The ovule is much enlarged where the stalk bends. The upper part of the ovule is on the whole like that of the pine, but there are two integuments (I, II) instead of the single one found in the pine.

As the seed develops, the embryo sac (*G, sp.*) enlarges so as to occupy pretty much the whole space of the seed. At first it is nearly filled with a fluid, but a layer of cells is formed, lining the walls, and this thickens until the whole space, except what is occupied by the small embryo, is filled with them. These are called the "endosperm cells," but differ from the endosperm cells of the gymnosperms, in the fact that they are not developed until after fertilization, and can hardly, therefore, be regarded as representing the prothallium of the gymnosperms and pteridophytes. These cells finally form a firm tissue, whose cells are filled with starch that forms a reserve supply of food for the embryo plant when the seed germinates. The embryo (Fig. 81, *K, em.*, Fig. 82, *J*), even when the seed is ripe, remains very small, and shows scarcely any differentiation. It is a small, pear-shaped mass of cells, the smaller end directed toward the upper end of the embryo sac.

The integuments grow with the embryo sac, and become brown and hard, forming the shell of the seed. The stalk of the ovule also enlarges, and finally forms the peculiar, spongy appendage of the seeds already noticed (Fig. 81, J, K).

CHAPTER XVI.
CLASSIFICATION OF THE MONOCOTYLEDONS.

IN the following chapter no attempt will be made to give an exhaustive account of the characteristics of each division of the monocotyledons, but only such of the most important ones as may serve to supplement our study of the special one already examined. The classification here, and this is the case throughout the spermaphytes, is based mainly upon the characters of the flowers and fruits.

The classification adopted here is that of the German botanist Eichler, and seems to the author to accord better with our present knowledge of the relationships of the groups than do the systems that are more general in this country. According to Eichler's classification, the monocotyledons may be divided into seven groups; viz., I. *Liliifloræ*; II. *Enantioblastæ*; III. *Spadicifloræ*; IV. *Glumaceæ*; V. *Scitamineæ*; VI. *Gynandræ*; VII. *Helobiæ*.

Order I. — *Liliifloræ*.

The plants of this group agree in their general structure with the adder's-tongue, which is a thoroughly typical representative of the group; but nevertheless, there is much variation among them in the details of structure. While most of them are herbaceous forms (dying down to the ground each year), a few, among which may be mentioned the yuccas ("bear grass," "Spanish bayonet") of our southern states, develop a creeping or upright woody stem, increasing in size from year to year. The herbaceous forms send up their stems yearly from underground bulbs, tubers, *e.g.* *Trillium* (Fig. 83, *A*), or thickened, creeping stems, or root stocks (rhizomes). Good examples of the last are the Solomon's-seal (Fig. 83, *B*), *Medeola* (*C*, *D*), and iris (Fig. 84 *A*). One family, the yams (*Dioscoreæ*), of which we have one common native species, the wild yam (*Dioscorea villosa*), have broad, netted-veined leaves and are twining plants, while another somewhat similar family (*Smilaceæ*) climb by means of tendrils at the bases of the leaves. Of the latter the "cat-brier" or "green-brier" is a familiar representative.

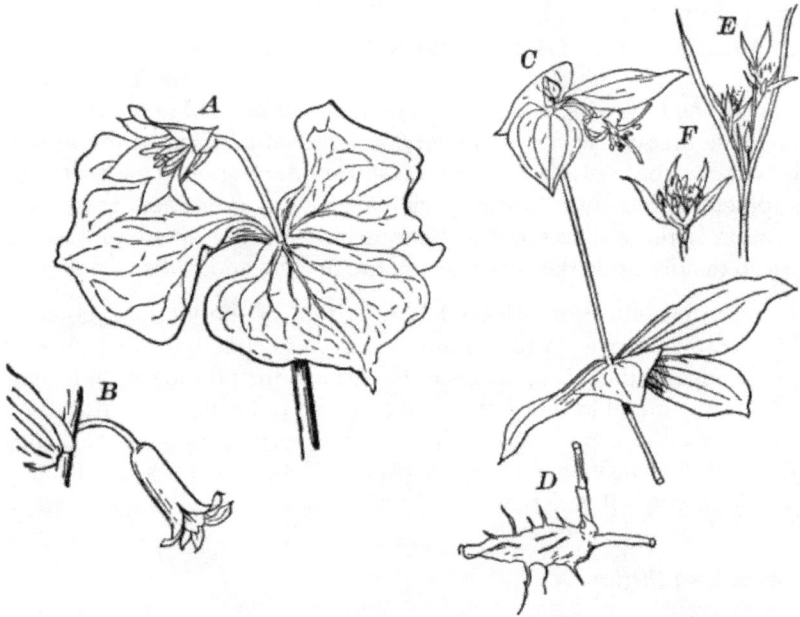

FIG. 83.—*Types of Liliiflora. A, Trillium,* × ¼. *B, single flower of Solomon's-seal (Polygonatum),* × 1. *C, upper part of a plant. D, underground stem (rhizome) of Indian cucumber root (Medeola),* × ½. *E, a rush (Juncus),* × 1. *F, a single flower,* × 2. *A–D, Liliaceæ; E, Juncaceæ.*

The flowers are for the most part conspicuous, and in plan like that of the adder's-tongue; but some, like the rushes (Fig. 83, *E*), have small, inconspicuous flowers; and others, like the yams and smilaxes, have flowers of two kinds, male and female.

FIG. 84.—*Types of Liliifloræ. A, flower of the common blue-flag (Iris),* × ½
*(Iridaceæ). B, the petal-like upper part of the pistil, seen from below, and
showing a stamen (an.). st. the stigma,* × ½. *C, the young fruit,* × ½. *D, section
of the same,* × 1. *E, diagram of the flower. F, part of a plant of the so-called
"gray moss" (Tillandsia),* × ½ *(Bromeliaceæ). G, a single flower,* × 2. *H, a
seed, showing the fine hairs attached to it,* × 1. *I, plant of pickerel-weed
(Pontederia),* × ¼ *(Pontederiaceæ). J, a single flower,* × 1. *K, section of the
ovary,* × 4.

The principal family of the *Liliifloræ* is the *Liliaceæ*, including
some of the most beautiful of all flowers. All of the true lilies (*Lilium*),
as well as the day lilies (*Funkia, Hemerocallis*) of the gardens, tulips,
hyacinths, lily-of-the-valley, etc., belong here, as well as a number of
showy wild flowers including several species of tiger-lilies (*Lilium*),
various species of *Trillium* (Fig. 83, *A*), Solomon's-seal (*Polygonatum*)
(Fig. 83, *B*), bellwort (*Uvularia*), and others. In all of these, except *Trillium*, the perigone leaves are colored alike, and the leaves parallel-
veined; but in the latter the sepals are green and the leaves broad and
netted-veined. The fruit of the *Liliaceæ* may be either a pod, like that
of the adder's-tongue, or a berry, like that of asparagus or Solomon's-
seal.

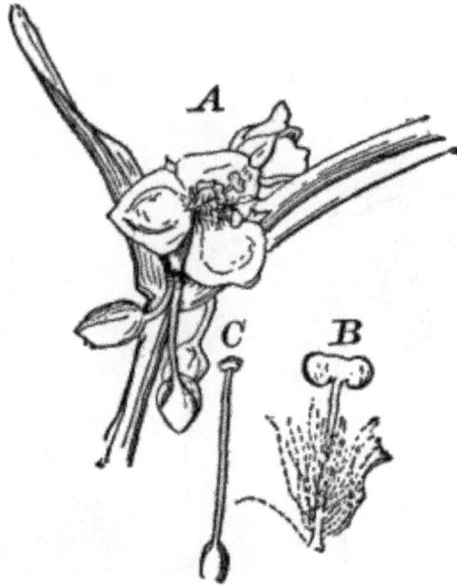

FIG. 85.—*Enantioblastæ. A, inflorescence of the common spiderwort (Trades-cantia),* × ½ *(Commelyneæ). B, a single stamen, showing the hairs attached to the filament,* × 2. *C, the pistil,* × 2.

Differing from the true lilies in having the bases of the perigone leaves adherent to the surface of the ovary, so that the latter is apparently below the flower (inferior), and lacking the inner circle of stamens, is the iris family (*Iridaceæ*), represented by the wild blue-flag (*Iris versicolor*) (Fig. 84, *A, E*), as well as by numerous cultivated species. In iris the carpels are free above and colored like the petals (*B*), with the stigma on the under side. Of garden flowers the gladiolus and crocus are the most familiar examples, besides the various species of iris; and of wild flowers the little "blue-eyed grass" (*Sisyrinchium*).

The blue pickerel-weed (*Pontederia*) is the type of a family of which there are few common representatives (Fig. 84, *I, K*).

The last family of the order is the *Bromeliaceæ*, all inhabitants of the warmer parts of the globe, but represented in the southern states by several forms, the commonest of which is the so-called "gray moss" (*Tillandsia*) (Fig. 84, *F, H*). Of cultivated plants the pineapple, whose fruit consists of a fleshy mass made up of the crowded fruits and the fleshy flower stalks, is the best known.

Order II. – *Enantioblastæ.*

The second order of the monocotyledons, *Enantioblastæ*, includes very few common plants. The most familiar examples are the various species of *Tradescantia* (Fig. 88), some of which are native, others exotic. Of the cultivated forms the commonest is one sometimes called "wandering-jew," a trailing plant with zigzag stems, and oval, pointed leaves forming a sheath about each joint. Another common one is the spiderwort already referred to. In this the leaves are long and pointed, but also sheathing at the base. When the flowers are showy, as in these, the sepals and petals are different, the former being green. The flowers usually open but once, and the petals shrivel up as the flower fades. There are four families of the order, the spiderwort belonging to the highest one, *Commelyneæ.*

Order III. – *Spadicifloræ.*

The third order of the monocotyledons, *Spadicifloræ*, is a very large one, and includes the largest and the smallest plants of the whole sub-class. In all of them the flowers are small and often very inconspicuous; usually, though not always, the male and female flowers are separate, and often on different plants. The smallest members of the group are little aquatics, scarcely visible to the naked eye, and of extremely simple structure, but nevertheless these little plants produce true flowers. In marked contrast to these are the palms, some of which reach a height of thirty metres or more.

The flowers in most of the order are small and inconspicuous, but aggregated on a spike (spadix) which may be of very large size. Good types of the order are the various aroids (*Aroideæ*), of which the calla (*Richardia*) is a very familiar cultivated example. Of wild forms the sweet-flag (*Acorus*), Jack-in-the-pulpit (*Arisæma*) (Fig. 86, *A, D*), skunk-cabbage (*Symplocarpus*), and wild calla may be noted. In *Arisæma* (Fig. 86, *A*) the flowers are borne only on the base of the spadix, and the plant is diœcious. The flowers are of the simplest structure, the female consisting of a single carpel, and the male of four stamens (*C, D*). While the individual flowers are destitute of a perigone, the whole inflorescence (cluster of flowers) is surrounded by a large leaf (spathe), which sometimes is brilliantly colored, this serving to attract insects. The leaves of the aroids are generally large and sometimes compound, the only instance of true compound leaves among the monocotyledons (Fig. 86, *B*).

FIG. 86.—*Types of Spadiciflorae. A, inflorescence of Jack-in-the-pulpit (Arisæma, Aroideæ). The flowers (fl.) are at the base of a spike (spadix), surrounded by a sheath (spathe), which has been cut away on one side in order to show the flowers, × ½. B, leaf of the same plant, × ¼. C, vertical section of a female flower, × 2. D, three male flowers, each consisting of four stamens, × 2. E, two plants of a duck-weed (Lemna), the one at the left is in flower, × 4. F, another common species. L, Trisulea, × 1. G, male flower of E, × 25. H, optical section of the female flower, showing the single ovule (ov.), × 25. I, part of the inflorescence of the bur-reed (Sparganium), with female flowers, × ½ (Typhaceæ). J, a single, female flower, × 2. K, a ripe fruit, × 1. L, longitudinal section of the same. M, two male flowers, × 1. N, a pond-weed (Potomogeton), × 1 (Naiadaceæ). O, a single flower, × 2. P, the same, with the perianth removed, × 2. Q, fruit of the same, × 2.*

Probably to be regarded as reduced aroids are the duck-weeds (*Lemnaceæ*) (Fig. 86, *F, H*), minute floating plants without any differentiation of the plant body into stem and leaves. They are globular or discoid masses of cells, most of them having roots; but one genus

179

(*Wolffia*) has no roots nor any trace of fibro-vascular bundles. The flowers are reduced to a single stamen or carpel (Figs. *E, G, H*).

The cat-tail (*Typha*) and bur-reed (*Sparganium*) (Fig. 86, *I, L*) are common representatives of the family *Typhaceæ*, and the pond-weeds (*Naias* and *Potomogeton*) are common examples of the family *Naiadeæ*. These are aquatic plants, completely submerged (*Naias*), or sometimes partially floating (*Potomogeton*). The latter genus includes a number of species with leaves varying from linear (very narrow and pointed) to broadly oval, and are everywhere common in slow streams.

The largest members of the group are the screw-pines (*Pandaneæ*) and the palms (*Palmæ*). These are represented in the United States by only a few species of the latter family, confined to the southern and southwestern portions. The palmettoes (*Sabal* and *Chamærops*) are the best known.

Both the palms and screw-pines are often cultivated for orna-ment, and as is well known, in the warmer parts of the world the palms are among the most valuable of all plants. The date palm (*Phœnix dactylifera*) and the cocoanut (*Cocos nucifera*) are the best known. The apparently compound ("pinnate" or feather-shaped) leaves of many palms are not strictly compound; that is, they do not arise from the branching of an originally single leaf, but are really broad, undivided leaves, which are closely folded like a fan in the bud, and tear apart along the folds as the leaf opens.

Although these plants reach such a great size, an examination of the stem shows that it is built on much the same plan as that of the other monocotyledons; that is, the stem is composed of a mass of soft, ground tissue through which run many small isolated, fibro-vascular bundles. A good idea of this structure may be had by cutting across a corn-stalk, which is built on precisely the same pattern.

Order IV. — *Glumaceæ*.

The plants of this order resemble each other closely in their habit, all having long, narrow leaves with sheathing bases that surround the slender, distinctly jointed stem which frequently has a hard, polished surface. The flowers are inconspicuous, borne usually in close spikes, and destitute of a perigone or having this reduced to small scales or hairs. The flowers are usually surrounded by more or less dry leaves (glumes, paleæ) which are closely set, so as to nearly conceal the flowers. The flowers are either hermaphrodite or unisexual.

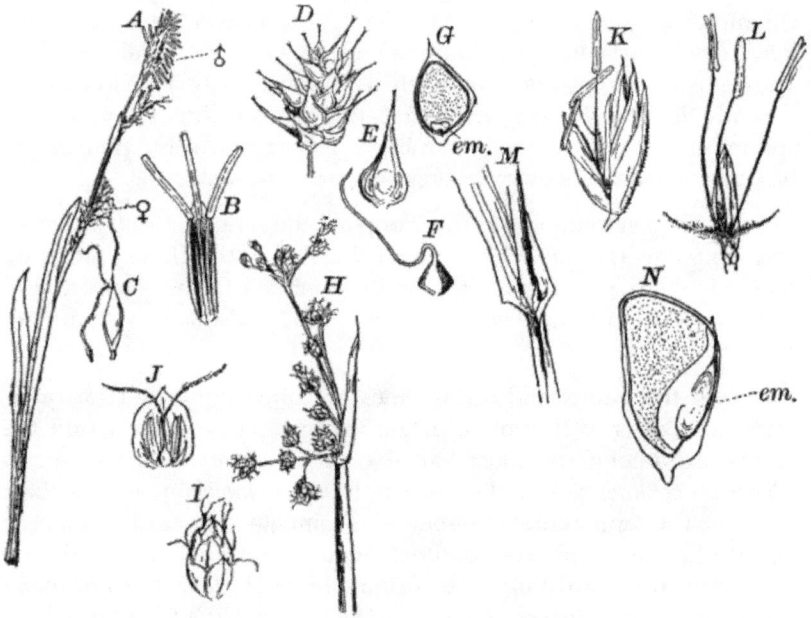

FIG. 87.—*Types of Glumaceæ. A, a sedge, Carex (Cyperaceæ). ♂, the male; ♀, the female flowers, × ½. B, a single male flower, × 2. C, a female flower, × 2. D, fruiting spike of another Carex, × ½. E, a single fruit, × 1. F, the same, with the outer envelope removed, and slightly enlarged. G, section of F, × 3. em. the embryo. H, a bulrush, Scirpus (Cyperaceæ), × ½. I, a single spikelet, × 2. J, a single flower, × 3. K, a spikelet of flowers of the common orchard grass, Dactylis (Gramineæ), × 2. L, a single flower, × 2. M, the base of a leaf, showing the split sheath encircling the stem, × 1. N, section of a kernel of corn, showing the embryo (em.), × 2.*

There are two well-marked families, the sedges (*Cyperaceæ*) and the grasses (*Gramineæ*). The former have solid, often triangular stems, and the sheath at the base of the leaves is not split. The commonest genera are *Carex* (Fig. 87, *A*, *G*) and *Cyperus*, of which there are many common species, differing very little and hard to distinguish. There are several common species of *Carex* which blossom early in the spring, the male flowers being quite conspicuous on account of the large, yellow anthers. The female flowers are in similar spikes lower down, where the pollen readily falls upon them, and is caught by the

181

long stigmas. In some other genera, *e.g.* the bulrushes (*Scirpus*) (Fig. 87, *H*), the flowers are hermaphrodite, *i.e.* contain both stamens and pistils. The fruit (Fig. 87, *F*) is seed-like, but really includes the wall of the ovary as well, which is grown closely to the enclosed seed. The embryo is small, surrounded by abundant endosperm (Fig. 87, *G*). Very few of the sedges are of any economic importance, though one, the papyrus of Egypt, was formerly much valued for its pith, which was manufactured into paper.

The second family, the grasses, on the contrary, includes the most important of all food plants, all of the grains belonging here. They differ mainly from the sedges in having, generally, hollow, cylindrical stems, and the sheath of the leaves split down one side; the leaves are in two rows, while those of the sedges are in three. The flowers (Fig. 87, *L*) are usually perfect; the stigmas, two in number and like plumes, so that they readily catch the pollen which is blown upon them. A few, like the Indian corn, have the flowers unisexual; the male flowers are at the top of the stem forming the "tassel," and the female flowers lower down forming the ear. The "silk" is composed of the enormously lengthened stigmas. The fruits resemble those of the sedges, but the embryo is usually larger and placed at one side of the endosperm (*N, em.*).

While most of the grasses are comparatively small plants, a few of them are almost tree-like in their proportions, the species of bamboo (*Bambusa*) sometimes reaching a height of twenty to thirty metres, with stems thirty to forty centimetres in diameter.

Order V. — *Scitamineæ.*

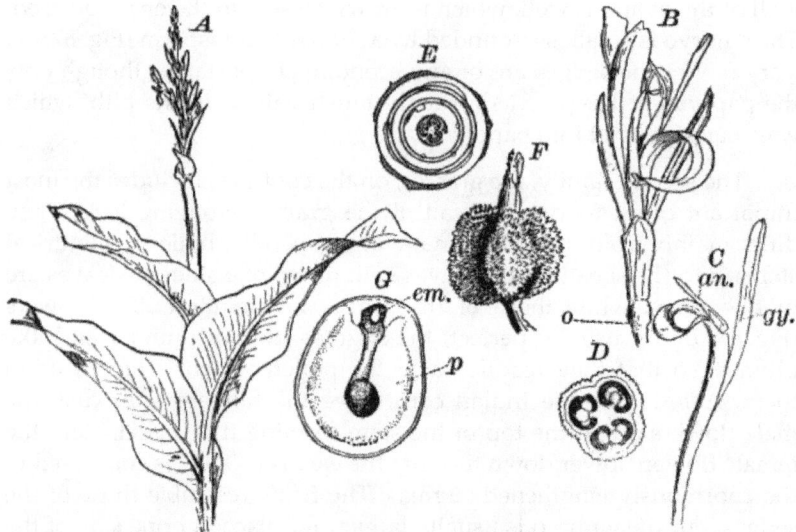

FIG. 88.—*Scitamineæ. A, upper part of a flowering plant of Indian shot (Canna), much reduced in size (Cannaceæ). B, a single flower, × ½. C, the single stamen (an.), and petal-like pistil (gy.), × 1. D, section of the ovary, × 2. E, diagram of the flower. The place of the missing stamens is indicated by small circles. F, fruit, × ½. G, section of an unripe seed. em. embryo. p, per-isperm, × 2.*

The plants of this order are all inhabitants of the warmer parts of the earth, and only a very few occur within the limits of the United States, and these confined to the extreme south. They are extremely showy plants, owing to their large leaves and brilliant flowers, and for this reason are cultivated extensively. Various species of *Canna* (Fig. 88) are common in gardens, where they are prized for their large, richly-colored leaves, and clusters of scarlet, orange, or yellow flow-ers. The leafy stems arise from thick tubers or root stocks, and grow rapidly to a height of two metres or more in the larger species. The leaves, as in all the order, are very large, and have a thick midrib with lateral veins running to the margin. The young leaves are folded up like a trumpet. The flowers are irregular in form, and in *Canna* only a single stamen is found; or if more are present, they are reduced to petal-like rudiments. The single, perfect stamen (Fig. 88, *C, an.*) has the

183

filament broad and colored like the petals, and the anther attached to one side. The pistil (*gy.*) is also petal-like. There are three circles of leaves forming the perigone, the two outer being more or less membranaceous, and only the three inner petal-like in texture. The ovary (*o*) is inferior, and covered on the outside with little papillæ that afterward form short spines on the outside of the fruit (*F*).

The seeds are large, but the embryo is very small. A section of a nearly ripe seed shows the embryo (*em.*) occupying the upper part of the embryo sac which does not nearly fill the seed and contains no endosperm. The bulk of the seed is derived from the tissue of the body of the ovule, which in most seeds becomes entirely obliterated by the growth of the embryo sac. The cells of this tissue become filled with starch, and serve the same purpose as the endosperm of other seeds. This tissue is called "perisperm."

Of food plants belonging to this order, the banana (*Musa*) is much the most important. Others of more or less value are species of arrowroot (*Maranta*) and ginger (*Zingiber*).

There are three families: I. *Musaceæ* (banana family); II. *Zingiberaceæ* (ginger family); and III. *Cannaceæ* (*Canna, Maranta*).

Order VI. — *Gynandræ*.

By far the greater number of the plants of this order belong to the orchis family (*Orchideæ*), the second family of the order (*Apostasieæ*), being a small one and unrepresented in the United States. The orchids are in some respects the most highly specialized of all flowers, and exhibit wonderful variety in the shape and color of the flowers, which are often of extraordinary beauty, and show special contrivances for cross-fertilization that are without parallel among flowering plants.

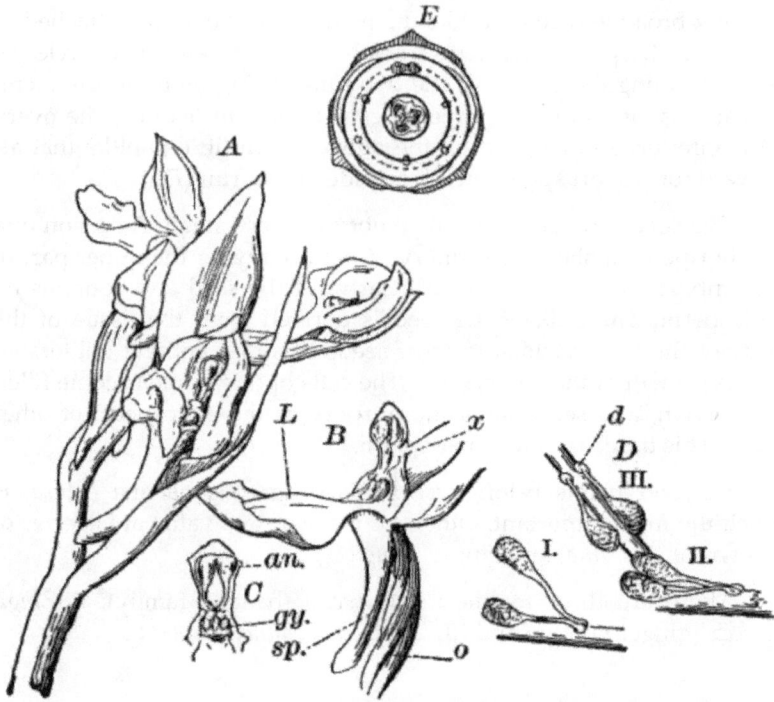

FIG. 89. — *Gynandræ*. *A, inflorescence of the showy orchis (Orchis spectabilis),* × 1 *(Orchideæ). B, a single flower, with the upper leaves of the perianth turned back to show the column (x). sp. the spur attached to the lower petal or lip. o, the ovary,* × 1. *C, the column seen from in front. an. the stamen. gy. the stigmatic surface,* × 1. *D, the two pollen masses attached to a straw, which was inserted into the flower, by means of the viscid disc (d): I, the masses immediately after their withdrawal; II, III, the same a few minutes later, showing the change in position. E, diagram of the flower; the position of the missing stamens indicated by small circles.*

The flowers are always more or less bilaterally symmetrical (zygomorphic). The ovary is inferior, and usually twisted so as to turn the flower completely around. There are two sets of perigone leaves, three in each, and these are usually much alike except the lower (through the twisting of the ovary) of the inner set. This petal, known as the "lip" or "labellum," is usually larger than the others, and different in color, as well as being frequently of peculiar shape. In many of them it is also prolonged backward in a hollow spur (see Fig. 89, *B*). In all of the orchids except the lady's-slippers (*Cypripedium*) (Fig. 90, *B*), only one perfect stamen is developed, and this is united with the three styles to form a special structure known, as the "column" or "gynostemium" (Fig. 89, B, C). The pollen spores are usually aggregated into two or four waxy masses ("pollinia," sing. pollinium), which usually can only be removed by the agency of insects upon which all but a very few orchids are absolutely dependent for the pollination of the flowers.

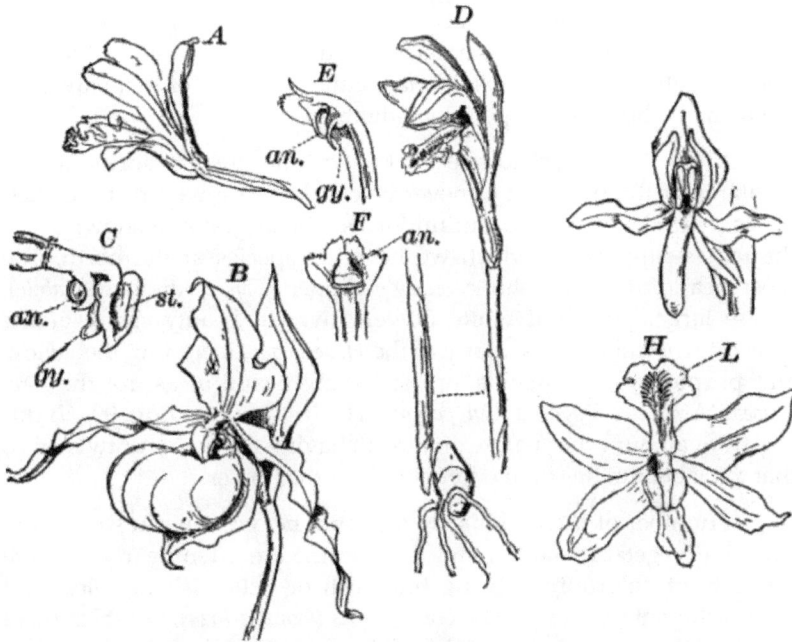

FIG. 90.—Forms of Orchideæ. A, putty-root (Aplectrum), × 1. B, yellow lady's-slipper (Cypripedium), × ½. C, the column of the same, × 1. an. one of the two perfect stamens. st. sterile, petal-like stamen. gy.. stigma. D, Arethusa, × ½. E, section of the column, × 1: an. stamen. gy. stigma. F, the same, seen from in front. G, Habenaria, × 1. H, Calopogon, × 1. In the last the ovary is not twisted, so that the lip (L) lies on the upper side of the flower.

In the lady-slippers there are two fertile stamens, and a third sterile one has the form of a large triangular shield terminating the column (Fig. 90, C, st.).

The ovules of the orchids are extremely small, and are only partly developed at the time the flower opens, the pollen tube growing very slowly and the ovules maturing as it grows down through the tissues of the column. The ripe seeds are excessively numerous, but so fine as to look like dust.

The orchids are mostly small or moderate-sized plants, few of them being more than a metre or so in height. All of our native species, with the exception of a few from the extreme south, grow from fibrous roots or tubers, but many tropical orchids, as is well known, are "epiphytes"; that is, they grow upon the trunks and branches of trees. One genus, *Vanilla*, is a twining epiphyte; the fruit of this plant furnishes the vanilla of commerce. Aside from this plant, the economical value of the orchids is small, although a few of them are used medicinally, but are not specially valuable.

Of the five thousand species known, the great majority are inhabitants of the tropics, but nevertheless there are within the United States a number of very beautiful forms. The largest and showiest are the lady's-slippers, of which we have six species at the north. The most beautiful is the showy lady's-slipper (*Cypripedium spectabile*), whose large, pink and white flowers rival in beauty many of the choicest tropical orchids. Many of the *Habenarias*, including the yellow and purple fringed orchids, are strikingly beautiful as are the *Arethuseæ* (*Arethusa, Pogonia, Calopogon*). The last of these (Fig. 90, *H*) differs from all our other native orchids in having the ovary untwisted so that the labellum lies on the upper side of the flower.

A number of the orchids are saprophytic, growing in soil rich in decaying vegetable matter, and these forms are often nearly or quite destitute of chlorophyll, being brownish or yellowish in color, and with rudimentary leaves. The coral roots (*Corallorhiza*), of which there are several species, are examples of these, and another closely related form, the putty-root (*Aplectrum*) (Fig. 90, *A*), has the flowering stems like those of *Corallorhiza*, but there is a single, large, plaited leaf sent up later.

Order VII. —*Helobiæ*.

The last order of the monocotyledons is composed of marsh or water plants, some of which recall certain of the dicotyledons. Of the three families, the first, *Juncagineæ*, includes a few inconspicuous plants with grass-like or rush-like leaves, and small, greenish or yellowish flowers (*e.g.* arrow-grass, *Triglochin*).

187

The second family (*Alismaceæ*) contains several large and showy species, inhabitants of marshes. Of these the water-plantain (*Alisma*), a plant with long-stalked, oval, ribbed leaves, and a much-branched panicle of small, white flowers, is very common in marshes and ditches, and the various species of arrowhead (*Sagittaria*) are among the most characteristic of our marsh plants. The flowers are unisexual; the female flowers are usually borne at the base of the inflorescence, and the male flowers above. The gynœcium (Fig. 91, *B*) consists of numerous, separate carpels attached to a globular receptacle. The sepals are green and much smaller than the white petals. The leaves (*F*) are broad, and, besides the thickened, parallel veins, have numerous smaller ones connecting these.

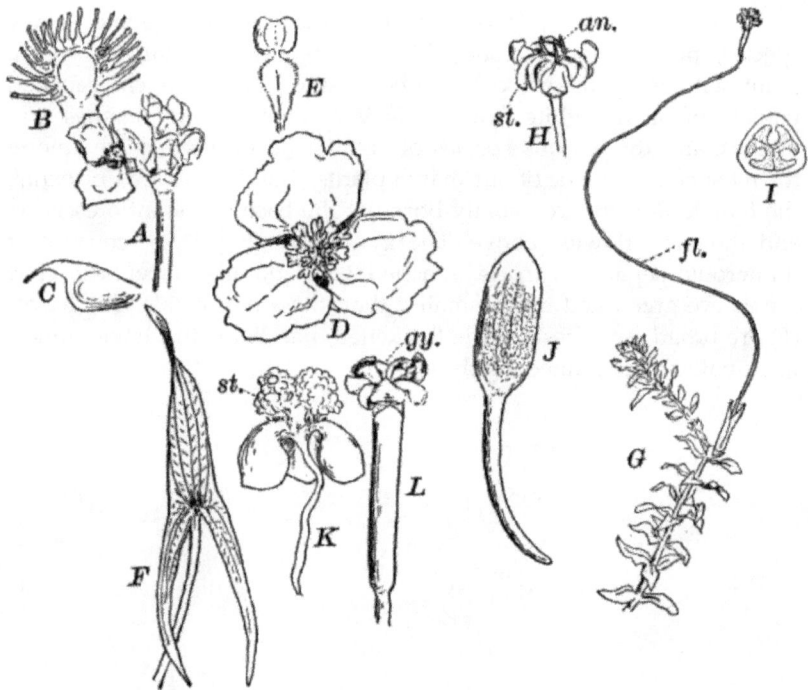

Fig. 91. — *Types of Helobiæ. A, inflorescence of arrowhead (Sagittaria), with a single female flower, × ½ (Alismaceæ). B, section through the gynœcium, showing the numerous single carpels, × 3. C, a ripe fruit, × 3. D, a male flower, × 1. E, a single stamen, × 3. F, a leaf of Sagittaria variabilis, × ¼. G, ditch-moss (Elodea), with a female flower (fl.), × ½. (Hydrocharideæ). H, the flower, × 2. an. the rudimentary stamens. st. the stigma. I, cross-section of the ovary, × 4. J, male inflorescence of eel-grass (Vallisneria), × 1. K, a single expanded male flower, × 12. st. the stamen. L, a female flower, × 1. gy. the stigma.*

The last family is the *Hydrocharideæ.* They are submersed aquatics, or a few of them with long-stalked, floating leaves. Two forms, the ditch-moss (*Elodea*) (Fig. 91, *G, I*) and eel-grass (*Vallisneria*) are very common in stagnant or slow-running water. In both of these the plants are completely submersed, but there is a special arrangement for bringing the flowers to the surface of the water. Like the arrow-head, the flowers are unisexual, but borne on different plants. The female flowers (*H, L*) are comparatively large, especially in *Vallisneria*, and are borne on long stalks, by means of which they reach the surface of the water, where they expand and are ready for pollination. The male flowers (Fig. 91, *J, K*) are extremely small and borne, many together, surrounded by a membranous envelope, the whole inflorescence attached by a short stalk. When the flowers are ready to open, they break away from their attachment, and the envelope opens, allowing them to escape, and they immediately rise to the surface where they expand and collect in great numbers about the open female flowers. Sometimes these are so abundant during the flowering period (late in summer) that the surface of the water looks as if flour had been scattered over it. After pollination is effected, the stem of the female flower coils up like a spring, drawing the flower beneath the water where the fruit ripens.

The cells of these plants show very beautifully the circulation of the protoplasm, the movement being very marked and continuing for a long time under the microscope. To see this the whole leaf of *Elodea*, or a section of that of *Vallisneria*, may be used.

CHAPTER XVII.
DICOTYLEDONS.

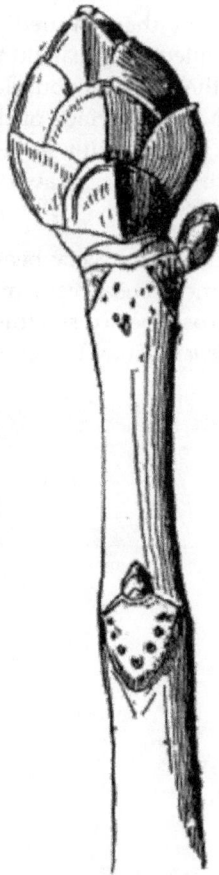

FIG. 92.—*End of a branch of a horsechestnut in winter, showing the buds covered by the thick, brown scale leaves,* × 1.

THE second sub-class of the angiosperms, the dicotyledons, receive their name from the two opposite seed leaves or cotyledons with which the young plant is furnished. These leaves are usually quite different in shape from the other leaves, and not infrequently are very thick and fleshy, filling nearly the whole seed, as may be seen in a bean or pea. The number of the dicotyledons is very large, and very much the greater number of living spermaphytes belong to this group.

They exhibit much greater variety in the structure of the flowers than the monocotyledons, and the leaves, which in the latter are with few exceptions quite uniform in structure, show here almost infinite variety. Thus the leaves may be simple (undivided); *e.g.* oak, apple; or compound, as in clover, locust, rose, columbine, etc. The leaves may be stalked or sessile (attached directly to the stem), or even grown around the stem, as in some honeysuckles. The edges of the leaves may be perfectly smooth ("entire"), or they may be variously lobed, notched, or wavy in many ways. As many of the dicotyledons are trees or shrubs that lose their leaves annually, special leaves are developed for the protection of the young leaves during the winter. These have the form of thick scales, and often are provided with glands secreting a gummy substance which helps render them waterproof. These scales are best studied in trees with large, winter buds, such as the horsechestnut (Fig. 92), hickory, lilac, etc. On removing the hard, scale leaves, the delicate, young leaves, and often the flowers, may be found within the bud. If we examine a young shoot of lilac or buckeye, just as the leaves are expanding in the spring, a complete series of forms may be seen from the simple, external scales, through immediate forms, to the complete foliage leaf. The veins of the leaves are almost always much-branched, the veins either being given off from one main vein or midrib (feather-veined or pinnate-veined), as in an apple leaf, or there may be a number of large veins radiating from the base of the leaf, as in the scarlet geranium or mallow. Such leaves are said to be palmately veined.

Some of them are small herbaceous plants, either upright or prostrate upon the ground, over which they may creep extensively, becoming rooted at intervals, as in the white clover, or sending out special runners, as is seen in the strawberry. Others are woody stemmed plants, persisting from year to year, and often becoming great trees that live for hundreds of years. Still others are climbing plants, either twining their stems about the support, like the morning-glory, hop, honeysuckle, and many others, or having special organs (tendrils) by which they fasten themselves to the support. These tendrils originate in different ways. Sometimes, as in the grape and Virginia creeper, they are reduced branches, either coiling about the support, or producing little suckers at their tips by which they cling to walls or the trunks of trees. Other tendrils, as in the poison ivy and the true ivy, are short roots that fasten themselves firmly in the crevices of bark or stones. Still other tendrils, as those of the sweet-pea and clematis, are parts of the leaf.

The stems may be modified into thorns for protection, as we see in many trees and shrubs, and parts of leaves may be similarly changed, as in the thistle. The underground stems often become much changed, forming bulbs, tubers, root stocks, etc. much as in the mono-cotyledons. These structures are especially found in plants which die down to the ground each year, and contain supplies of nourishment for the rapid growth of the annual shoots.

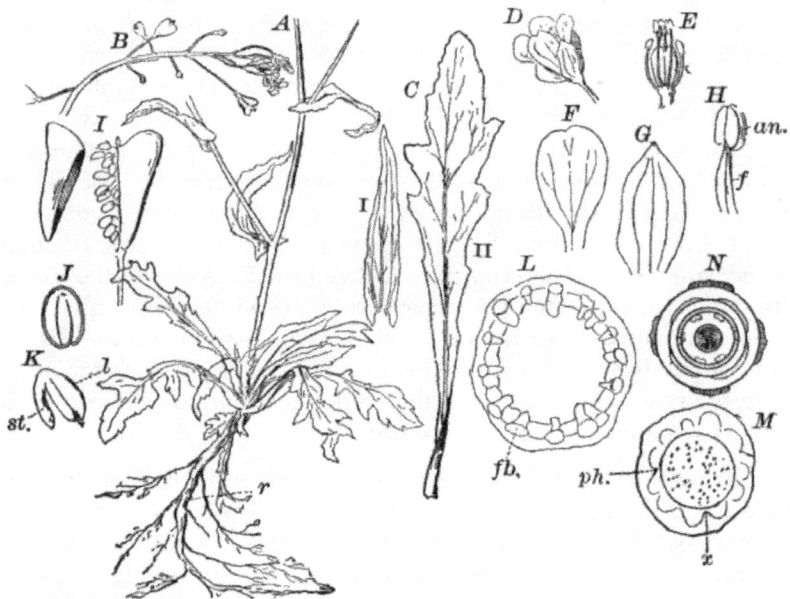

FIG. 93.—A, base of a plant of shepherd's-purse (Capsella bursa-pastoris), × ½. r, the main root. B, upper part of the inflorescence, × 1. C, two leaves: I, from the upper part; II, from the base of the plant, × 1. D, a flower, × 3. E, the same, with sepals and petals removed, × 3. F, petal. G, sepal. H, stamen, × 10. f, filament. an. anther. I, a fruit with one of the valves removed to show the seeds, × 4. J, longitudinal section of a seed, × 8. K, the embryo removed from the seed, × 8. l, the first leaves (cotyledons). st. the stem ending in the root. L, cross-section of the stem, × 20. fb. fibro-vascular bundle. M, a similar section of the main root, × 15. N, diagram of the flower.

The structure of the tissues, and the peculiarities of the flower and fruit, will be better understood by a somewhat careful examination of a typical dicotyledon, and a comparison with this of examples of the principal orders and families.

One of the commonest of weeds, and at the same time one of the most convenient plants for studying the characteristics of the dicotyledons, is the common shepherd's-purse (*Capsella bursa-pastoris*) (Figs. 93–95).

The plant grows abundantly in waste places, and is in flower nearly the year round, sometimes being found in flower in midwinter, after a week or two of warm weather. It is, however, in best condition for study in the spring and early summer. The plant may at once be recognized by the heart-shaped pods and small, white, four-petaled flowers. The plant begins to flower when very small, but continues to grow until it forms a much-branching plant, half a metre or more in height. On pulling up the plant, a large tap-root (Fig. 93, *A*, *r*) is seen, continuous with the main stem above ground. The first root of the seedling plant continues here as the main root of the plant, as was the case with the gymnosperms, but not with the monocotyledons. From this tap-root other small ones branch off, and these divide repeatedly, forming a complex root system. The main root is very tough and hard, owing to the formation of woody tissue in it. A cross-section slightly magnified (Fig. 93, *M*), shows a round, opaque, white, central area (*x*), the wood, surrounded by a more transparent, irregular ring (*ph.*), the phloem or bast; and outside of this is the ground tissue and epidermis.

The lower leaves are crowded into a rosette, and are larger than those higher up, from which they differ also in having a stalk (petiole), while the upper leaves are sessile. The outline of the leaves varies much in different plants and in different parts of the same plant, being sometimes almost entire, sometimes divided into lobes almost to the midrib, and between these extremes all gradations are found. The larger leaves are traversed by a strong midrib projecting strongly on the lower side of the leaf, and from this the smaller veins branch. The upper leaves have frequently two smaller veins starting from the base of the leaf, and nearly parallel with the midrib (*C* I). The surface of the leaves is somewhat roughened with hairs, some of which, if slightly magnified, look like little white stars.

Magnifying slightly a thin cross-section of the stem, it shows a central, ground tissue (pith), whose cells are large enough to be seen even when very slightly enlarged. Surrounding this is a ring of fibro-vascular bundles (*L*, *fb*.), appearing white and opaque, and connected by a more transparent tissue. Outside of the ring of fibro-vascular bundles is the green ground tissue and epidermis. Comparing this with the section of the seedling pine stem, a resemblance is at once evident, and this arrangement was also noticed in the stem of the horse-tail.

Branches are given off from the main stem, arising at the point where the leaves join the stem (axils of the leaves), and these may in turn branch. All the branches terminate finally in an elongated inflorescence, and the separate flowers are attached to the main axis of the inflorescence by short stalks. This form of inflorescence is known technically as a "raceme." Each flower is really a short branch from which the floral leaves arise in precisely the same way as the foliage leaves do from the ordinary branches. There are five sets of floral leaves: I. four outer perigone leaves (sepals) (*F*), small, green, pointed leaves traversed by three simple veins, and together forming the calyx; II. four larger, white, inner perigone leaves (petals) (*G*), broad and slightly notched at the end, and tapering to the point of attachment. The petals collectively are known as the "corolla." The veins of the petals fork once; III. and IV. two sets of stamens (*E*), the outer containing two short, and the inner, four longer ones arranged in pairs. Each stamen has a slender filament (*H*, *f*) and a two-lobed anther (*an*.). The innermost set consists of two carpels united into a compound pistil. The ovary is oblong, slightly flattened so as to be oval in section, and divided into two chambers. The style is very short and tipped by a round, flattened stigma.

The raceme continues to grow for a long time, forming new flowers at the end, so that all stages of flowers and fruit may often be found in the same inflorescence.

The flowers are probably quite independent of insect aid in pollination, as the stamens are so placed as to almost infallibly shed their pollen upon the stigma. This fact, probably, accounts for the inconspicuous character of the flowers.

After fertilization is effected, and the outer floral leaves fall off, the ovary rapidly enlarges, and becomes heart-shaped and much flattened at right angles to the partition. When ripe, each half falls away, leaving the seeds attached by delicate stalks (funiculi, sing. funiculus) to the edges of the membranous partition. The seeds are small, oval bodies with a shining, yellow-brown shell, and with a little dent at the end where the stalk is attached. Carefully dividing the seed length-wise, or crushing it in water so as to remove the embryo, we find it occupies the whole cavity of the seed, the young stalk (*st.*) being bent down against the back of one of the cotyledons (*f*).

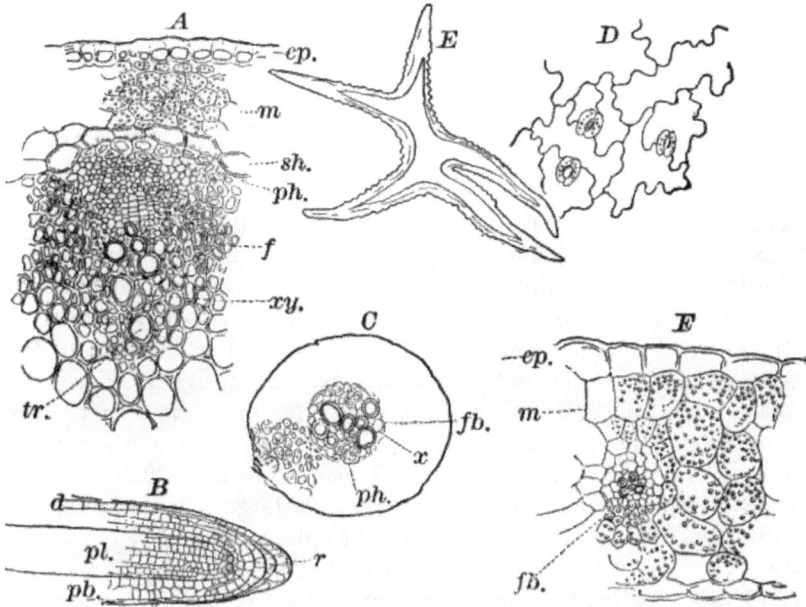

FIG. 94.—*A, cross-section of the stem of the shepherd's-purse, including a fibro-vascular bundle, × 150. ep. epidermis. m, ground tissue. sh. bundle sheath. ph. phloem. xy. xylem. tr. a vessel. B, a young root seen in optical section, × 150. r, root cap. d, young epidermis. pb. ground. pl. young fibro-vascular bundle. C cross section of a small root, × 150. fb. fibro-vascular bundle. D, epidermis from the lower side of the leaf, × 150. E, a star-shaped hair from the surface of the leaf, × 150. F, cross-section of a leaf, × 150. ep. epidermis. m, ground tissue. fb. section of a vein.*

A microscopic examination of a cross-section of the older root shows that the central portion is made up of radiating lines of thick-walled cells (fibres) interspersed with lines of larger, round openings (vessels). There is a ring of small cambium cells around this merging into the phloem, which is composed of irregular cells, with pretty thick, but soft walls. The ground tissue is composed of large, loose cells, which in the older roots are often ruptured and partly dried up.

The epidermis is usually indistinguishable in the older roots. To understand the early structure of the roots, the smallest rootlets obtainable should be selected. The smallest are so transparent that the tips may be mounted whole in water, and will show very satisfactorily the arrangement of the young tissues. The tissues do not here arise from a single, apical cell, as we found in the pteridophytes, but from a group of cells (the shaded cells in Fig. 94, B). The end of the root, as in the fern, is covered with a root cap (r) composed of successive layers of cells cut off from the growing point. The rest of the root shows the same division of the tissues into the primary epidermis (dermatogen) (d), young fibro-vascular cylinder (plerome) (pl.), and young ground tissue (periblem) (pb.). The structure of the older portions of such a root is not very easy to study, owing to difficulty in making good cross-sections of so small an object. By using a very sharp razor, and holding perfectly straight between pieces of pith, however, satisfactory sections can be made. The cells contain so much starch as to make them almost opaque, and potash should be used to clear them. The fibro-vascular bundle is of the radial type, there being two masses of xylem (xy.) joined in the middle, and separating the two phloem masses (ph.), some of whose cells are rather thicker walled than the others. The bundle sheath is not so plain here as in the fern. The ground tissue is composed of comparatively large cells with thickish, soft walls, that contain much starch. The epidermis usually dies while the root is still young. In the larger roots the early formation of the cambium ring, and the irregular arrangement of the tissues derived from its growth, soon obliterate all traces of the primitive arrangement of the tissues. Making a thin cross-section of the stem, and magnifying strongly, we find bounding the section a single row of epidermal cells (Fig. 94, A, ep.) whose walls, especially the outer ones, are strongly thickened. Within these are several rows of thin-walled ground-tissue cells containing numerous small, round chloroplasts. The innermost row of these cells (sh.) are larger and have but little chlorophyll. This row of cells forms a sheath around the ring of fibro-vascular bundles very much as is the case in the horse-tail. The separate bundles are nearly triangular in outline, the point turned inward, and are connected with each other by masses of fibrous tissue (f), whose thickened walls have a peculiar, silvery lustre. Just inside of the bundle sheath there is a row of similar fibres marking the outer limit of the phloem (ph.). The rest of the phloem is composed of very small cells. The xylem is composed of fibrous cells with yellowish walls and numerous large vessels (tr.). The central ground tissue (pith) has large, thin-walled cells with numerous intercellular spaces, as in the stem of *Erythronium*. Some of these cells contain a few scattered chloroplasts in the very thin, protoplasmic layer lining their walls, but the cells are almost completely filled with colorless cell sap.

A longitudinal section shows that the epidermal cells are much elongated, the cells of the ground tissue less so, and in both the partition walls are straight. In the fibrous cells, both of the fibro-vascular bundle and those lying between, the end walls are strongly oblique. The tracheary tissue of the xylem is made up of small, spirally-marked vessels, and larger ones with thickened rings or with pits in the walls. The small, spirally-marked vessels are nearest the centre, and are the first to be formed in the young bundle.

The epidermis of the leaves is composed of irregular cells with wavy outlines like those of the ferns. Breathing pores, of the same type as those in the ferns and monocotyledons, are found on both surfaces, but more abundant and more perfectly developed on the lower surface of the leaf. Owing to their small size they are not specially favorable for study. The epidermis is sparingly covered with unicellular hairs, some of which are curiously branched, being irregularly star-shaped. The walls of these cells are very thick, and have little protuberances upon the outer surface (Fig. 93, *E*).

Cross-sections of the leaf may be made between pith as already directed; or, by folding the leaf carefully several times, the whole can be easily sectioned. The structure is essentially as in the adder-tongue, but the epidermal cells appear more irregular, and the fibro-vascular bundles are better developed. They are like those of the stem, but somewhat simpler. The xylem lies on the upper side.

The ground tissue is composed, as in the leaves we have studied, of chlorophyll-bearing, loose cells, rather more compact upon the upper side. (In the majority of dicotyledons the upper surface of the leaves is nearly or quite destitute of breathing pores, and the cells of the ground tissue below the upper epidermis are closely packed, forming what is called the "palisade-parenchyma" of the leaf.)

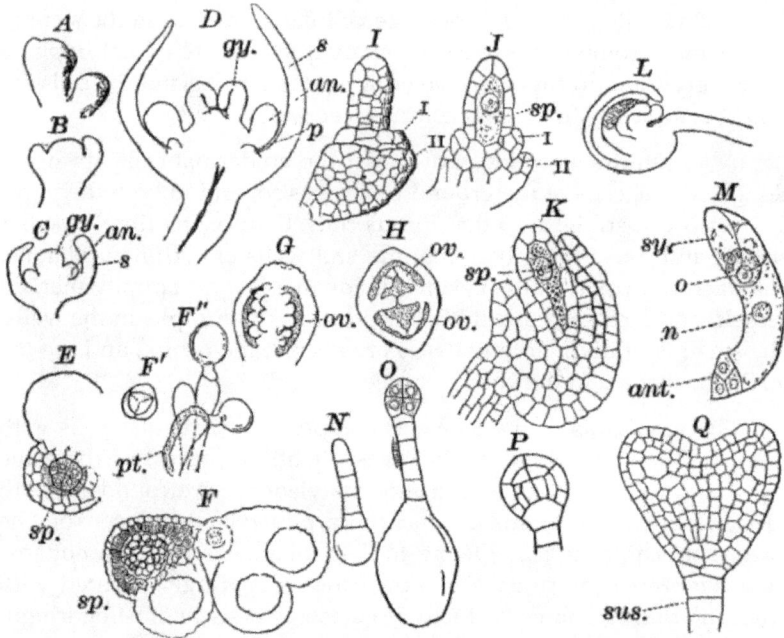

FIG. 95.—A–D, *successive stages in the development of the flower of Capsella,* × 50. *A, surface view. B–D, optical sections. s, sepals, p, petals. an. stamens. gy. pistil. E, cross-section of the young anther, × 180. sp. spore mother cells. F, cross-section of full-grown anther. sp. pollen spores, × 50. F□, four young pollen spores, × 300. F□, pollen spores germinating upon the stigma, × 300. pt. pollen tube. G, young pistil in optical section, × 25. H, cross-section of a somewhat older one. ov. ovules. I–L, development of the ovule. sp. embryo sac (macrospore). I–K, × 150. L, × 50. M, embryo sac of a full-grown ovule, × 150. Sy. Synergidæ. o, egg cell. n, endosperm nucleus. ant. antipodal cells. N–Q, development of the embryo, × 150. sus. suspensor.*

The shepherd's-purse is an admirable plant for the study of the development of the flower which is much the same in other angio-sperms. To study this, it is only necessary to teaze out, in a drop of water, the tip of a raceme, and putting on a cover glass, examine with a power of from fifty to a hundred diameters. In the older stages it is best to treat with potash, which will render the young flowers quite transparent. The young flower (Fig. 95, *A*) is at first a little protuber-ance composed of perfectly similar small cells filled with dense proto-plasm. The first of the floral leaves to appear are the sepals which very early arise as four little buds surrounding the young flower axis (Fig. 95, *A, B*). The stamens (*C, an.*) next appear, being at first entirely

199

similar to the young sepals. The petals do not appear until the other parts of the flower have reached some size, and the first tracheary tissue appears in the fibro-vascular bundle of the flower stalk (*D*). The carpels are more or less united from the first, and form at first a sort of shallow cup with the edges turned in (*D, gy*.). This cup rapidly elongates, and the cavity enlarges, becoming completely closed at the top where the short style and stigma develop. The ovules arise in two lines on the inner face of each carpel, and the tissue which bears them (placenta) grows out into the cavity of the ovary until the two placentæ meet in the middle and form a partition completely across the ovary (Fig. 95, *H*).

The stamens soon show the differentiation into filament and anther, but the former remains very short until immediately before the flowers are ready to open. The anther develops four sporangia (pollen sacs), the process being very similar to that in such pteridophytes as the club mosses. Each sporangium (Fig. *E, F*) contains a central mass of spore mother cells, and a wall of three layers of cells. The spore mother cells finally separate, and the inner layer of the wall cells becomes absorbed much as we saw in the fern, and the mass of mother cells thus floats free in the cavity of the sporangium. Each one now divides in precisely the same way as in the ferns and gymnosperms, into four pollen spores. The anther opens as described for *Erythronium*.

By carefully picking to pieces the young ovaries, ovules in all stages of development may be found, and on account of their small size and transparency, show beautifully their structure. Being perfectly transparent, it is only necessary to mount them in water and cover.

The young ovule (*I, J*) consists of a central, elongated body (nucellus), having a single layer of cells enclosing a large central cell (the macrospore or embryo sac) (*sp*.). The base of the nucellus is surrounded by two circular ridges (I, II) of which the inner is at first higher than the outer one, but later (*K, L*), the latter grows up above it and completely conceals it as well as the nucellus. One side of the ovule grows much faster than the other, so that it is completely bent upon itself, and the opening between the integuments is brought close to the base of the ovule (Fig. 95, *L*). This opening is called the "micropyle," and allows the pollen tube to enter.

The full-grown embryo sac shows the same structure as that already described in *Monotropa* (page 276), but as the walls of the full-grown ovule are thicker here, its structure is rather difficult to make out. The ripe stigma is covered with little papillæ (Fig. 95, *F*) that hold the pollen spores which may be found here sending out the pollen tube. By carefully opening the ovary and slightly crushing it in a drop of water, the pollen tube may sometimes be seen growing along the stalk of the ovule until it reaches and enters the micropyle.

To study the embryo a series of young fruits should be selected, and the ovules carefully dissected out and mounted in water, to which a little caustic potash has been added. The ovule will be thus rendered transparent, and by pressing gently on the cover glass with a needle so as to flatten the ovule slightly, there is usually no trouble in seeing the embryo lying in the upper part of the embryo sac, and by pressing more firmly it can often be forced out upon the slide. The potash should now be removed as completely as possible with blotting paper, and pure water run under the cover glass.

The fertilized egg cell first secretes a membrane, and then divides into a row of cells (*N*) of which the one nearest the micropyle is often much enlarged. The cell at the other end next enlarges and becomes divided by walls at right angles to each other into eight cells. This globular mass of cells, together with the cell next to it, is the embryo plant, the row of cells to which it is attached taking no further part in the process, and being known as the "suspensor." Later the embryo becomes indented above and forms two lobes (*Q*), which are the beginnings of the cotyledons. The first root and the stem arise from the cells next the suspensor.

CHAPTER XVIII.
CLASSIFICATION OF DICOTYLEDONS.

Division I. — *Choripetalæ.*

NEARLY all of the dicotyledons may be placed in one of two great divisions distinguished by the character of the petals. In the first group, called *Choripetalæ*, the petals are separate, or in some degenerate forms entirely absent. As familiar examples of this group, we may select the buttercup, rose, pink, and many others.

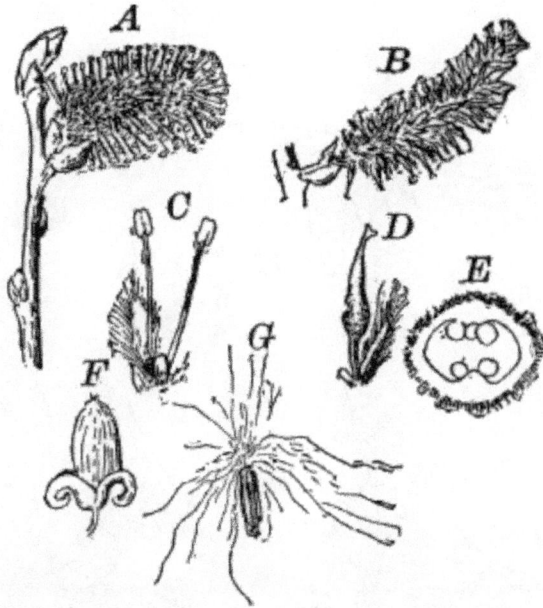

FIG. 96. — *Iuliflora. A, male; B, female inflorescence of a willow, Salix (Amentaceæ),* × ½. *C, a single male flower,* × 2. *D, a female flower,* × 2. *E, cross-section of the ovary,* × 8. *F, an opening fruit. G, single seed with its hairy appendage,* × 2.

The second group (*Sympetalæ* or *Gamopetalæ*) comprises those dicotyledons whose flowers have the petals more or less completely united into a tube. The honeysuckles, mints, huckleberry, lilac, etc., are familiar representatives of the *Sympetalæ*, which includes the highest of all plants.

The *Choripetalæ* may be divided into six groups, including twenty-two orders. The first group is called *Iuliflorœ*, and contains numerous, familiar plants, mostly trees. In these plants, the flowers are small and inconspicuous, and usually crowded into dense catkins, as in willows (Fig. 96) and poplars, or in spikes or heads, as in the lizard-tail (Fig. 97, *G*), or hop (Fig. 97, *I*). The individual flowers are very small and simple in structure, being often reduced to the gynœcium or andræcium, carpels and stamens being almost always in separate flowers. The outer leaves of the flower (sepals and petals) are either entirely wanting or much reduced, and never differentiated into calyx and corolla.

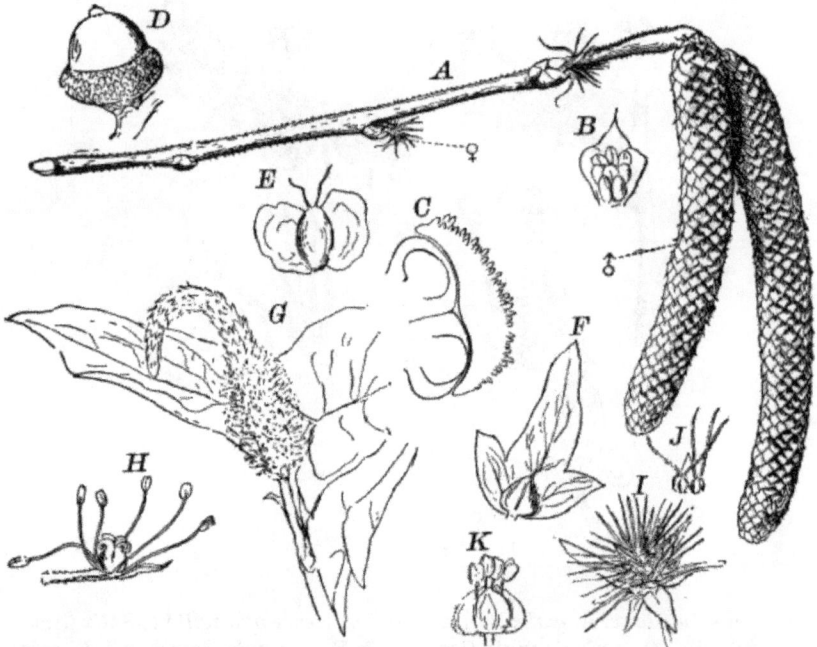

Fig. 97.—Types of Iuliflorœ. A, branch of hazel, Corylus (Cupuliferœ), × 1. ♂, male; ♀, female inflorescence. B, a single male flower, × 3. C, section of the ovary of a female flower, × 25. D, acorn of red oak, Quercus (Cupuliferœ), × ½. E, seed of white birch, Betula (Betulaceœ), × 3. F, fruit of horn-bean, Carpinus (Cupuliferœ), × 1. G, lizard-tail, Saururus (Saurureœ), × ¼. H, a single flower, × 2. I, female inflorescence of the hop, Humulus (Cannabineœ), × 1. J, a single scale with two flowers, × 1. K, a male flower of a nettle, Urtica (Urticaceœ), × 5.

203

In the willows (Fig. 96) the stamens are bright-colored, so that the flowers are quite showy, and attract numerous insects which visit them for pollen and nectar, and serve to carry the pollen to the pistillate flowers, thus insuring their fertilization. In the majority of the group, however, the flowers are wind-fertilized. An excellent example of this is seen in the common hazel (Fig. 97, A). The male flowers are produced in great numbers in drooping catkins at the ends of the branches, shedding the pollen in early spring before the leaves unfold. The female flowers are produced on the same branches, but lower down, and in much smaller numbers. The stigmas are long, and covered with minute hairs that catch the pollen which is shaken out in clouds every time the plant is shaken by the wind, and falls in a shower over the stigmas. A similar arrangement is seen in the oaks, hickories, and walnuts.

There are three orders of the *Iulifloræ*: *Amentaceæ*, *Piperineæ*, and *Urticinæ*. The first contains the birches (*Betulaceæ*); oaks, beeches, hazels, etc. (*Cupuliferæ*); walnuts and hickories (*Juglandeæ*); willows and poplars (*Salicaceæ*). They are all trees or shrubs; the fruit is often a nut, and the embryo is very large, completely filling it.

The *Piperineæ* are mostly tropical plants, and include the pepper plant (*Piper*), as well as other plants with similar properties. Of our native forms, the only common one is the lizard-tail (*Saururus*), not uncommon in swampy ground. In these plants, the calyx and corolla are entirely absent, but the flowers have both carpels and stamens (Fig. 97, H).

The *Urticinæ* include, among our common plants, the nettle family (*Urticaceæ*); plane family (*Plataneæ*), represented by the sycamore or buttonwood (*Platanus*); the hemp family (*Cannabineæ*); and the elm family (*Ulmaceæ*). The flowers usually have a calyx, and may have only stamens or carpels, or both. Sometimes the part of the stem bearing the flowers may become enlarged and juicy, forming a fruit-like structure. Well-known examples of this are the fig and mulberry.

The second group of the *Choripetalæ* is called *Centrospermæ*, and includes but a single order comprising seven families, all of which, except one (*Nyctagineæ*), are represented by numerous native species. The latter comprises mostly tropical plants, and is represented in our gardens by the showy "four-o'clock" (*Mirabilis*). In this plant, as in most of the order, the corolla is absent, but here the calyx is large and brightly colored, resembling closely the corolla of a morning-glory or petunia. The stamens are usually more numerous than the sepals, and the pistil, though composed of several carpels, has, as a rule, but a single cavity with the ovules arising from the base, though sometimes the ovary is several celled.

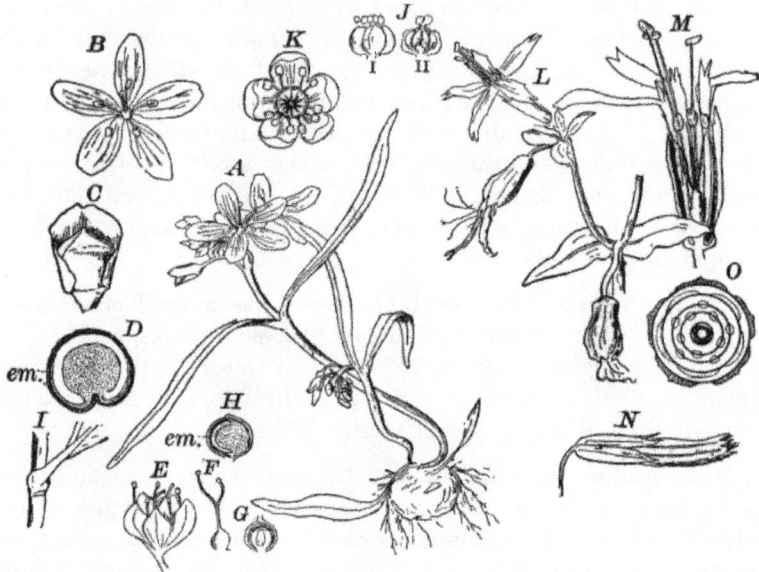

FIG. 98. — *Types of Centrospermæ. A, plant of spring-beauty, Claytonia (Portulacaceæ), × ½. B, a single flower, × 1. C, fruit, with the sepals removed, × 2. D, section of the seed, showing the curved embryo (em.), × 5. E, single flower of smart-weed, Polygonum (Polygonaceæ), × 2. F, the pistil, × 2. G, section of the ovary, showing the single ovule, × 4. H, section of the seed, × 2. I, base of the leaf, showing the sheath, × 1. J, flower of pig-weed, Chenopodium (Chenopodiaceæ), × 3: I, from without; II, in section. K, flower of the poke-weed, Phytolacca (Phytolaccaceæ), × 2. L, fire-pink, Silene (Caryophyllaceæ), × ½. M, a flower with half of the calyx and corolla removed, × 1. N, ripe fruit of mouse-ear chick-weed, Cerastium (Caryophyllaceæ), opening by ten teeth at the summit, × 2. O, diagram of the flower of Silene.*

The first family (*Polygoneæ*) is represented by the various species of *Polygonum* (knotgrass, smart-weed, etc.), and among cultivated plants by the buckwheat (*Fagopyrum*). The goose-foot or pig-weed (*Chenopodium*) among native plants, and the beet and spinach of the gardens are examples of the family *Chenopodiaceæ*. Nearly resembling the last is the amaranth family (*Amarantaceæ*), of which the showy amaranths and coxcombs of the gardens, and the coarse, green amaranth or pig-weed are representatives.

The poke-weed (*Phytolacca*) (Fig. 98, *K*), so conspicuous in autumn on account of its dark-purple clusters of berries and crimson stalks, is our only representative of the family *Phytolaccaceæ*. The two highest families are the purslane family (*Portulaceæ*) and pink family (*Caryophylleæ*). These are mostly plants with showy flowers in which the petals are large and conspicuous, though some of the pink family, *e.g.* some chick-weeds, have no petals. Of the purslane family the portulacas of the gardens, and the common purslane or "pusley," and the spring-beauty (*Claytonia*) (Fig. 98, *A*) are the commonest examples. The pink family is represented by many common and often showy plants. The carnation, Japanese pinks, and sweet-william, all belonging to the genus *Dianthus*, of which there are also two or three native species, are among the showiest of the family. The genera *Lychnis* and *Silene* (Fig. 98, *L*) also contain very showy species. Of the less conspicuous genera, the chick-weeds (*Cerastium* and *Stellaria*) are the most familiar.

The third group of the *Choripetalæ* (the *Aphanocyclæ*) is a very large one and includes many common plants distributed among five orders. The lower ones have all the parts of the flower entirely separate, and often indefinite in number; the higher have the gynœcium composed of two or more carpels united to form a compound pistil.

The first order (*Polycarpæ*) includes ten families, of which the buttercup family (*Ranunculaceæ*) is the most familiar. The plants of this family show much variation in the details of the flowers, which are usually showy, but the general plan is much the same. In some of them, like the anemones (Fig. 99, *A*), clematis, and others, the corolla is absent, but the sepals are large and brightly colored so as to appear like petals. In the columbine (*Aquilegia*) (Fig. 99, *F*) the petals are tubular, forming nectaries, and in the larkspur (Fig. 99, *T*) one of the sepals is similarly changed.

Representing the custard-apple family (*Anonaceæ*) is the curious papaw (*Asimina*), common in many parts of the United States (Fig. 100, *A*). The family is mainly a tropical one, but this species extends as far north as southern Michigan.

FIG. 99. — Types of Aphanocyclæ (Polycarpæ), family Ranunculaceæ. A, Rue anemone (Anemonilla), × ½. B, a fruit, × 2. C, section of the same. D, section of a buttercup flower (Ranunculus), × 1½. E, diagram of buttercup flower. F, wild columbine (Aquilegia), × ½. G, one of the spur-shaped petals, × 1. H, the five pistils, × 1. I, longitudinal section of the fruit, × 1. J, flower of larkspur (Delphinium), × 1. K, the four petals and stamens, after the removal of the five colored and petal-like sepals, × 1.

The magnolia family (Magnoliaceæ) has several common members, the most widely distributed being, perhaps, the tulip-tree (Liriodendron) (Fig. 100, C), much valued for its timber. Besides this there are several species of magnolia, the most northerly species being the sweet-bay (Magnolia glauca) of the Atlantic States, and the cucumbertree (M. acuminata); the great magnolia (M. grandiflora) is not hardy in the northern states.

The sweet-scented shrub (Calycanthus) (Fig. 100, G) is the only member of the family Calycanthaceæ found within our limits. It grows wild in the southern states, and is cultivated for its sweet-scented, dull, reddish flowers.

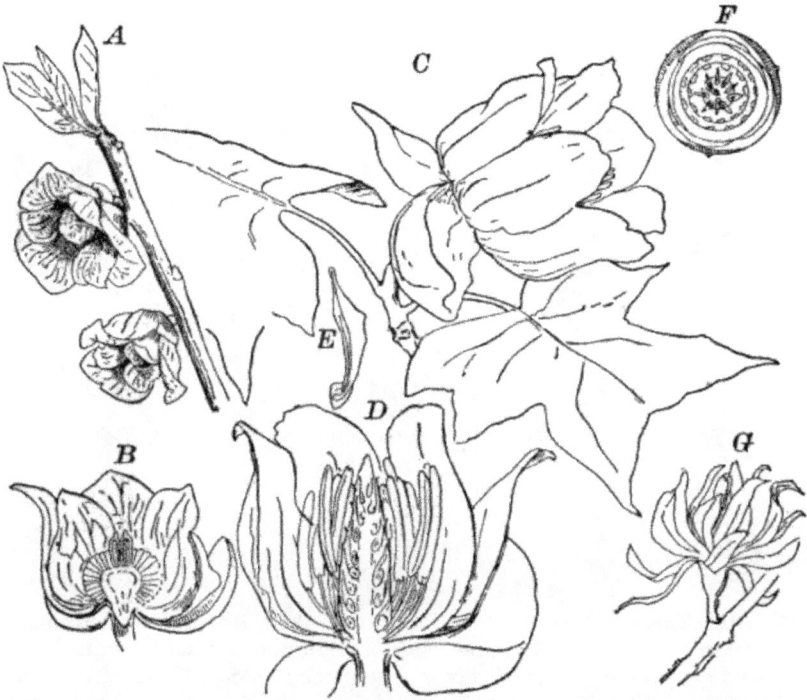

FIG. 100.—*Types of Aphanocyclæ (Polycarpæ). A, branch of papaw, Asimina (Anonaceæ),* × ½. *B, section of the flower,* × 1. *C, flower and leaf of tulip-tree, Liriodendron (Magnoliaceæ),* × ⅓. *D, section of a flower,* × ½. *E, a ripe fruit,* × 1. *F, diagram of the flower. G, flower of the sweet-scented shrub, Calycanthus (Calycanthaceæ),* × ½

The barberry (*Berberis*) (Fig. 101, *A*) is the type of the family *Berberideæ*, which also includes the curious mandrake or may-apple (*Podophyllum*) (Fig. 101, *D*), and the twin-leaf or rheumatism-root (*Jeffersonia*), whose curious seed vessel is shown in Figure 101, *G*. The fruit of the barberry and may-apple are edible, but the root of the latter is poisonous.

The curious woody twiner, moon-seed (*Menispermum*) (Fig. 101, *I*), is the sole example in the northern states of the family *Menispermeæ* to which it belongs. The flowers are diœcious, and the pistillate flowers are succeeded by black fruits looking like grapes. The flattened, bony seed is curiously sculptured, and has the embryo curled up within it.

208

FIG. 101. — *Types of Aphanocyclæ (Polycarpæ). A–H, Berberidaceæ. A, flower of barberry (Berberis), × 2. B, the same in section. C, a stamen, showing the method of opening, × 3. D, flower of may-apple (Podophyllum), × ½. E, section of the ovary of D, × 1. F, diagram of the flower. G, ripe fruit of twin-leaf (Jeffersonia), opening by a lid, × ½. H, section of seed, showing the embryo (em.), × 2. I, young leaf and cluster of male flowers of moon-seed, Menispermum (Menispermeæ), × 1. J, a single male flower, × 2. K, section of a female flower, × 2. L, ripe seed, × 1. M, section of L, showing the curved embryo.*

The last two families of the order, the laurel family (*Laurineæ*) and the nutmeg family (*Myristicineæ*) are mostly tropical plants, characterized by the fragrance of the bark, leaves, and fruit. The former is represented by the sassafras and spice-bush, common throughout the eastern United States. The latter has no members within our borders, but is familiar to all through the common nutmeg, which is the seed of *Myristica fragrans* of the East Indies. "Mace" is the "aril" or covering of the seed of the same plant.

The second order of the *Aphanocyclæ* comprises a number of aquatic plants, mostly of large size, and is known as the *Hydropeltidinæ*. The flowers and leaves are usually very large, the latter usually nearly round in outline, and frequently with the stalk inserted near the middle. The leaves of the perigone are numerous, and sometimes

209

merge gradually into the stamens, as we find in the common white water-lily (*Castalia*).

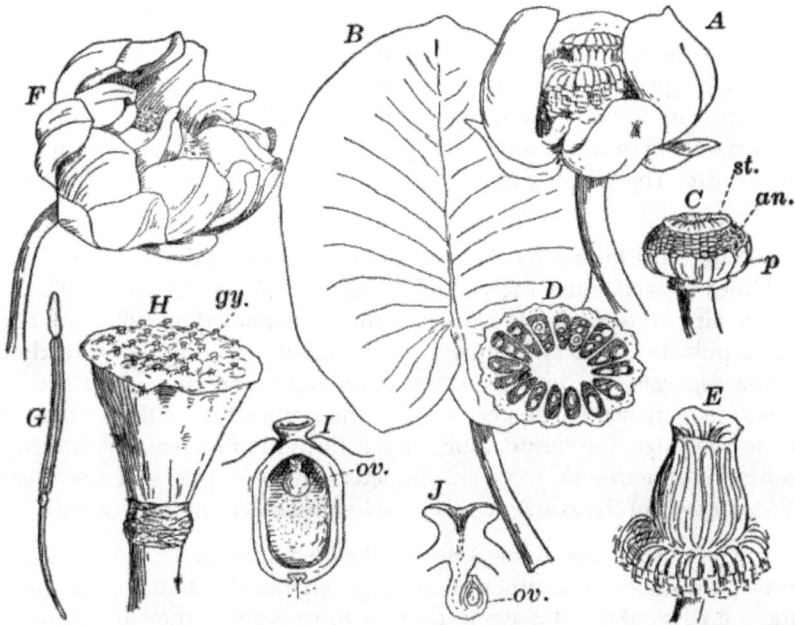

FIG. 102.—*Types of Aphanocyclæ (Hydropeltidinæ). A, yellow water-lily, Nymphæa (Nymphæaceæ), × ½. B, a leaf of the same, × ⅛. C, freshly opened flower, with the large petal-like sepals removed, × ½. p, petals. an. stamens. st. stigma. D, section of the ovary, × 2. E, young fruit, × ½. F, lotus, Nelumbo (Nelumbieæ). × ⅛. G, a stamen, × 1. H, the large receptacle, with the separate pistils sunk in its surface, × ½. I, section of a single pistil, × 2. ov. the ovule. J, upper part of a section through the stigma and ovule (ov.), × 4.*

There are three families, all represented within the United States. The first (*Nelumbieæ*) has but a single species, the yellow lotus or ne-lumbo (*Nelumbo lutea*), common in the waters of the west and south-west, but rare eastward (Fig. 101, *F*). In this flower, the end of the flower axis is much enlarged, looking like the rose of a watering-pot, and has the large, separate carpels embedded in its upper surface. When ripe, each forms a nut-like fruit which is edible. There are but two species of *Nelumbo* known, the second one (*N. speciosa*) being a native of southeastern Asia, and probably found in ancient times in Egypt, as it is represented frequently in the pictures and carvings of the ancient Egyptians. It differs mainly from our species in the color of its flowers which are red instead of yellow. It has recently been introduced into New Jersey where it has become well established in several localities.

The second family (*Cabombeæ*) is also represented at the north by but one species, the water shield (*Brasenia*), not uncommon in marshes. Its flowers are quite small, of a dull-purple color, and the leaves oval in outline and centrally peltate, *i.e.* the leaf stalk inserted in the centre. The whole plant is covered with a transparent gelatinous coat.

The third family (*Nymphæaceæ*) includes the common white water-lilies (*Castalia*) and the yellow water-lilies (*Nymphæa*) (Fig. 102, *A*). In the latter the petals are small and inconspicuous (Fig. 102, *C, p*), but the sepals are large and showy. In this family the carpels, instead of being separate, are united into a large compound pistil. The water-lilies reach their greatest perfection in the tropics, where they attain an enormous size, the white, blue, or red flowers of some species being thirty centimetres or more in diameter, and the leaves of the great *Victoria regia* of the Amazon reaching two metres or more in width.

The third order of the *Aphanocyclæ* (*Rhœadinæ* or *Crucifloræ*) comprises a number of common plants, principally characterized by having the parts of the flowers in twos or fours, so that they are more or less distinctly cross-shaped, whence the name *Cruciforæ*.

There are four families, of which the first is the poppy family (*Papaveraceæ*), including the poppies, eschscholtzias, Mexican or prickly poppy (*Argemone*), etc., of the gardens, and the blood-root (*Sanguinaria*), celandine poppy (*Stylophorum*), and a few other wild plants (see Fig. 103, *A–I*). Most of the family have a colored juice (latex), which is white in the poppy, yellow in celandine and *Argemone*, and orange-red in the blood-root. From the latex of the opium poppy the opium of commerce is extracted.

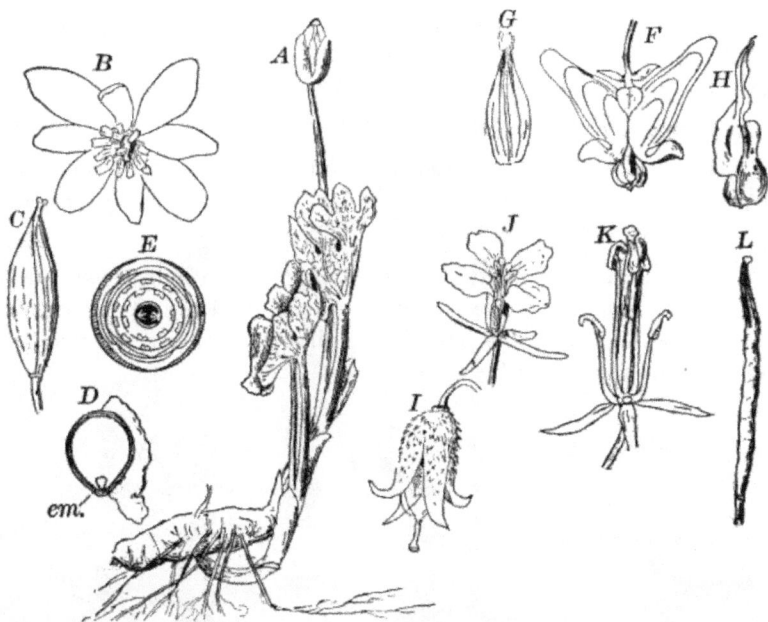

FIG. 103. — *Types of Aphanocyclæ (Rhœdinæ). A, plant of blood-root, Sanguinaria (Papaveraceæ),* × ⅓. *B, a single flower,* × 1. *C, fruit,* × ½. *D, section of the seed. em. embryo,* × 2. *E, diagram of the flower. F, flower of Dutchman's breeches, Dicentra (Fumariaceæ),* × 1. *G, group of three stamens of the same,* × 2. *H, one of the inner petals,* × 2. *I, fruit of celandine poppy, Stylophorum (Papaveraceæ),* × ½. *J, flower of mustard, Brassica (Cruciferæ),* × 1. *K, the same, with the petals removed,* × 2. *L, fruit of the same,* × 1.

The second family, the fumitories (*Fumariaceæ*) are delicate, smooth plants, with curious flowers and compound leaves. The garden bleeding-heart (*Dicentra spectabilis*) and the pretty, wild *Dicentras* (Fig. 103, *F*) are familiar to nearly every one.

Other examples are the mountain fringe (*Adlumia*), a climbing species, and several species of *Corydalis*, differing mainly from *Dicentra* in having the corolla one-sided.

The mustard family (*Cruciferæ*) comprises by far the greater part of the order. The shepherd's-purse, already studied, belongs here, and may be taken as a type of the family. There is great uniformity in all as regards the flowers, so that the classification is based mainly on differences in the fruit and seeds. Many of the most valuable garden vegetables, as well as a few more or less valuable wild plants, are members of the family, which, however, includes some troublesome weeds. Cabbages, turnips, radishes, with all their varieties, belong here, as well as numerous species of wild cresses. A few like the wallflower (*Cheiranthus*) and stock (*Matthiola*) are cultivated for ornament.

The last family is the caper family (*Capparideæ*), represented by only a few not common plants. The type of the order is *Capparis*, whose pickled flower-buds constitute capers.

The fourth order (*Cistifloræ*) of the *Aphanocyclæ* is a very large one, but the majority of the sixteen families included in it are not represented within our limits. The flowers have the sepals and petals in fives, the stamens either the same or more numerous.

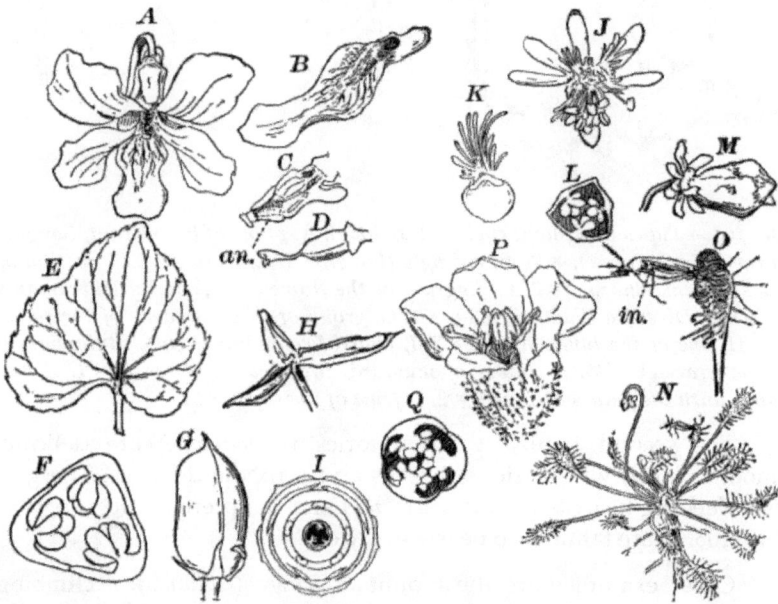

FIG. 104.—*Types of Aphanocyclæ (Cistifloræ). A, flower of wild blue violet, Viola (Violaceæ), × 1. B, the lower petal prolonged behind into a sac or spur, × 1. C, the stamens, × 2. D, pistil, × 2. E, a leaf, × ½. F, section of the ovary, × 2. G, the fruit, × 1. H, the same after it has opened, × 1. I, diagram of the flower. J, flower of mignonette, Reseda (Resedaceæ), × 2. K, a petal, × 3. L, cross-section of the ovary, × 3. M, fruit, × 1. N, plant of sundew, Drosera (Droseraceæ), × ½. O, a leaf that has captured a mosquito, × 2. P, flower of another species (D. filiformis), × 2. Q, cross-section of the ovary, × 4.*

Among the commoner members of the order are the mignonettes (*Resedaceæ*) and the violets (*Violaceæ*), of which the various wild and cultivated species are familiar plants (Fig. 104, *A, M*). The sundews (*Droseraceæ*) are most extraordinary plants, growing in boggy land over pretty much the whole world. They are represented in the United States by several species of sundew (*Drosera*), and the still more curious Venus's-flytrap (*Dionæa*) of North Carolina. The leaves of the latter are sensitive, and composed of two parts which snap together like a steel trap. If an insect lights upon the leaf, and touches certain hairs upon its upper surface, the two parts snap together, holding the insect tightly. A digestive fluid is secreted by glands upon the inner surface of the leaf, and in a short time the captured insect is actually digested and absorbed by the leaves. The same process takes place in the sundew (Fig. 104, *N*) where, however, the mechanism is somewhat different. Here the tentacles, with which the leaf is studded, secrete a sticky fluid which holds any small insect that may light upon it. The tentacles now slowly bend inward and finally the edges of the leaf as well, until the captured insect is firmly held, when a digestive process, similar to that in *Dionæa*, takes place. This curious habit is probably to be explained from the position where the plant grows, the roots being in water where there does not seem to be a sufficient supply of nitrogenous matter for the wants of the plant, which supplements the supply from the bodies of the captured insects.

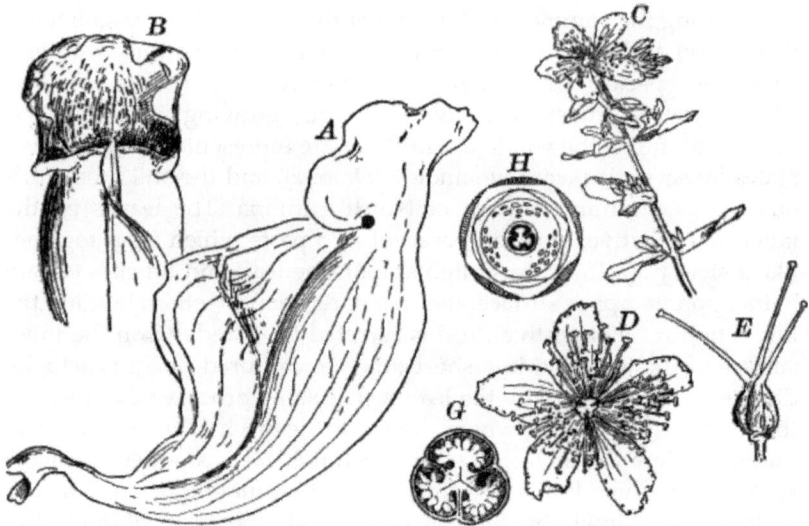

Fig. 105.—Types of Aphanocyclæ (Cistifloræ). A, B, leaves of the pitcher-plant, Sarracenia (Sarraceniaceæ). A, from the side; B, from in front, × ½. C, St. John's-wort (Hypericum), × ½. D, a flower, × 1. E, the pistil, × 2. G, cross-section of the ovary, × 4. H, diagram of the flower.

Similar in their habits, but differing much in appearance from the sundews, are the pitcher-plants (*Sarraceniaceæ*), of which one species (*Sarracenia purpurea*) is very common in peat bogs throughout the northern United States. In this species (Fig. 105, *A, B*), the leaves form a rosette, from the centre of which arises in early summer a tall stalk bearing a single, large, nodding, dark-reddish flower with a curious umbrella-shaped pistil. The leaf stalk is hollow and swollen, with a broad wing on one side, and the blade of the leaf forms a sort of hood at the top. The interior of the pitcher is covered above with stiff, downward-pointing hairs, while below it is very smooth. Insects readily enter the pitcher, but on attempting to get out, the smooth, slippery wall at the bottom, and the stiff, downward-directed hairs above, prevent their escape, and they fall into the fluid which fills the bottom of the cup and are drowned, the leaf absorbing the nitrogenous compounds given off during the process of decomposition. There are other species common in the southern states, and a California pitcher-plant (*Darlingtonia*) has a colored appendage at the mouth of the pitcher which serves to lure insects into the trap.

Another family of pitcher-plants (*Nepentheæ*) is found in the warmer parts of the old world, and some of them are occasionally cultivated in greenhouses. In these the pitchers are borne at the tips of the leaves attached to a long tendril.

Two other families of the order contain familiar native plants, the rock-rose family (*Cistaceæ*), and the St. John's-worts (*Hypericaceæ*). The latter particularly are common plants, with numerous showy yellow flowers, the petals usually marked with black specks, and the leaves having clear dots scattered through them. The stamens are numerous, and often in several distinct groups (Fig. 105, *C, D*).

The last order of the *Aphanocyclæ* (the *Columniferæ*) has three families, of which two, the mallows (*Malvaceæ*), and the lindens (*Tiliaceæ*), include well-known species. Of the former, the various species of mallows (Fig. 106, *A*) belonging to the genus *Malva* are common, as well as some species of *Hibiscus*, including the showy swamp *Hibiscus* or rose-mallow (*H. moscheutos*), common in salt marshes and in the fresh-water marshes of the great lake region. The hollyhock and shrubby *Althæa* are familiar cultivated plants of this order, and the cotton-plant (*Gossypium*) also belongs here. In all of these the stamens are much branched, and united into a tube enclosing the style. Most of them are characterized also by the development of great quantities of a mucilaginous matter within their tissues.

The common basswood (*Tilia*) is the commonest representative of the family *Tiliaceæ* (Fig. 106, *G*). The nearly related European linden, or lime-tree, is sometimes planted. Its leaves are ordinarily somewhat smaller than our native species, which it, however, closely resembles.

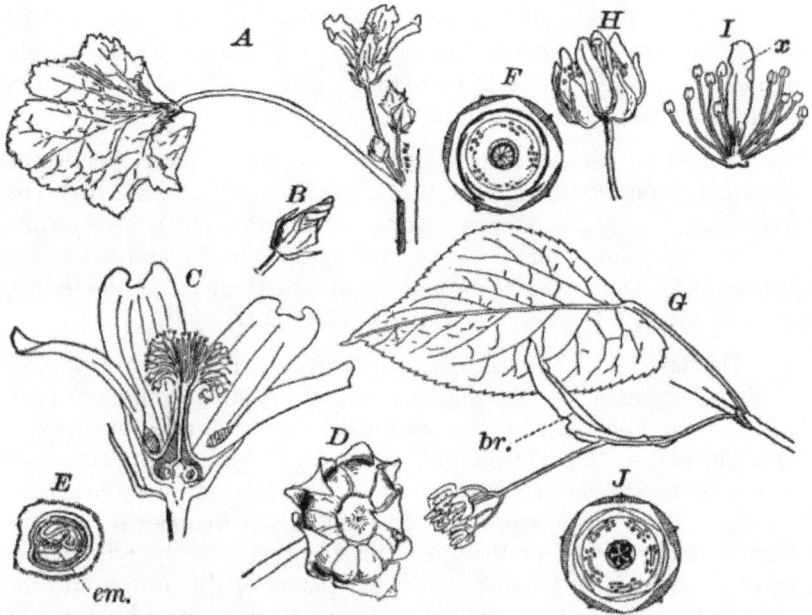

FIG. 106.—*Types of Aphanocyclæ (Columniferæ). A, flower and leaf of the common mallow, Malva (Malvaceæ), × ½. B, a flower bud, × 1. C, section of a flower, × 2. D, the fruit, × 2. E, section of one division of the fruit, with the enclosed seed, × 3. em. the embryo. F, diagram of the flower. G, leaf and inflorescence of the basswood, Tilia (Tiliaceæ), × ⅓. br. a bract. H, a single flower, × 1. I, group of stamens, with petal-like appendage (x), × 2. J, diagram of the flower.*

The fourth group of the *Choripetalæ* is the *Eucyclæ*. The flowers most commonly have the parts in fives, and the stamens are never more than twice as many as the sepals. The carpels are usually more or less completely united into a compound pistil. There are four orders, comprising twenty-five families.

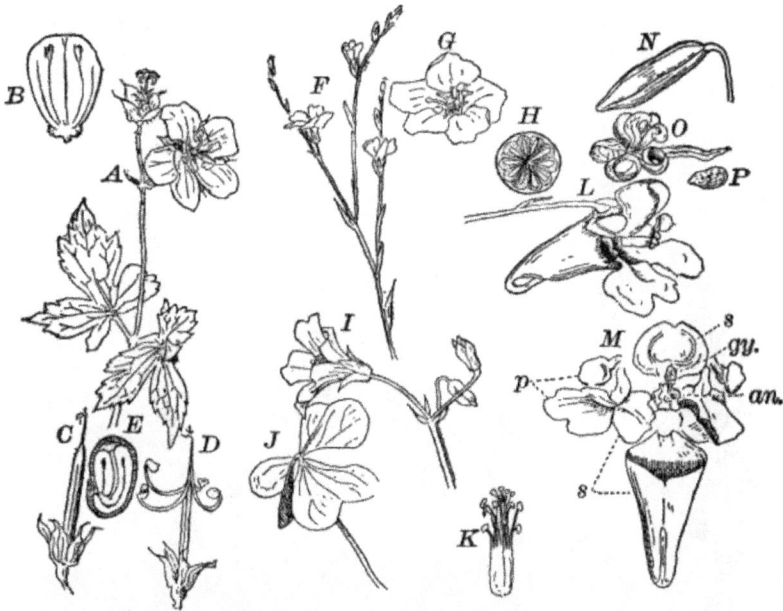

FIG. 107. — *Types of Eucyclæ (Gruinales). A, wild crane's-bill Geranium (Geraniaceæ)*, × ½. *B, a petal*, × 1. *C, the young fruit, the styles united in a column*, × ½. *D, the ripe fruit, the styles separating to discharge the seeds*, × ½. *E, section of a seed*, × 2. *F, wild flax. Linum (Linaceæ)*, × ½. *G, a single flower*, × 2. *H, cross-section of the young fruit*, × 3. *I, flower. J, leaf of wood-sorrel, Oxalis (Oxalideæ)*, × 1. *K, the stamens and pistil*, × 2. *L, flower of jewel-weed, Impatiens (Balsamineæ)*, × 1. *M, the same, with the parts separated. p, petals. s, sepals. an. stamens. gy. pistil. N, fruit*, × 1. *O, the same, opening. P, a seed*, × 2.

The first order (*Gruinales*) includes six families, consisting for the most part of plants with conspicuous flowers. Here belong the geraniums (Fig. 107, *A*), represented by the wild geraniums and crane's-bill, and the very showy geraniums (*Pelargonium*) of the gardens. The nasturtiums (*Tropæolum*) represent another family, mostly tropical, and the wood-sorrels (*Oxalis*) (Fig. 107, *I*) are common, both wild and cultivated. The most useful member of the order is unquestionably the common flax (*Linum*), of which there are also several native species (Fig. 107, *F*). These are types of the flax family (*Linaceæ*). Linen is the product of the tough, fibrous inner bark of *L. usitatissimum*, which has been cultivated for its fibre from time immemorial. The last family is the balsam family (*Balsamineæ*). The jewel-weed or touch-me-not (*Im-*

218

patiens), so called from the sensitive pods which spring open on being touched, is very common in moist ground everywhere (Fig. 107, *L–P*). The garden balsam, or lady's slipper, is a related species (*I. balsamina*).

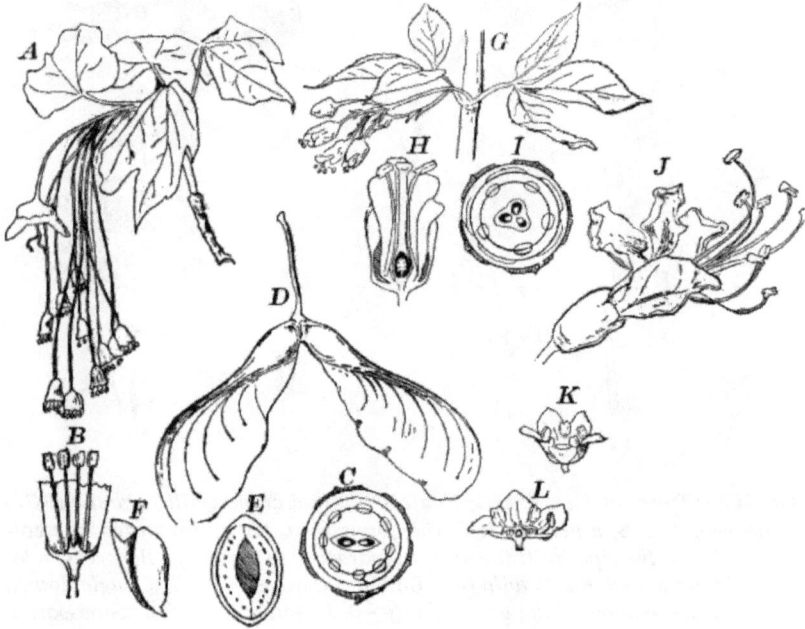

FIG. 108.—*Eucyclæ (Terebinthinæ, Æsculinæ). A, leaves and flowers of sugar-maple, Acer (Aceraceæ), × ½. B, a male flower, × 2. C, diagram of a perfect flower. D, fruit of the silver-maple, × ½. E, section across the seed, × 2. F, embryo removed from the seed, × 1. G, leaves and flowers of bladder-nut, Staphylea, (Sapindaceæ), × ½. H, section of a flower, × 2. I, diagram of the flower. J, flower of buckeye (Æsculus), × 1½. K, flower of smoke-tree, Rhus (Anacardiaceæ), × 3. L, the same, in section.*

The second order (*Terebinthinæ*) contains but few common plants. There are six families, mostly inhabitants of the warmer parts of the world. The best-known members of the order are the orange, lemon, citron, and their allies. Of our native plants the prickly ash (*Zanthoxylum*), and the various species of sumach (*Rhus*), are the best known. In the latter genus belong the poison ivy (*R. toxicodendron*) and the poison dogwood (*R. venenata*). The Venetian sumach or smoke-tree (*R. Cotinus*) is commonly planted for ornament.

The third order of the *Eucyclæ*, the *Æsculinæ*, embraces six families, of which three, the horsechestnuts, etc. (*Sapindaceæ*), the maples
(*Aceraceæ*), and the milkworts (*Polygalaceæ*), have several representatives in the northern United States. Of the first the buckeye (*Æsculus*)
(Fig. 108, *J*) and the bladder-nut (*Staphylea*) (Fig. 108, *G*) are the commonest native genera, while the horsechestnut (*Æsculus hippocasta-
num*) is everywhere planted.

The various species of maple (*Acer*) are familiar examples of the
Aceraceæ (see Fig. 106, *A*, *F*).

The fourth and last order of the *Eucyclæ*, the *Frangulinæ*, is composed mainly of plants with inconspicuous flowers, the stamens as
many as the petals. Not infrequently they are diœcious, or in some,
like the grape, some of the flowers may be unisexual while others are
hermaphrodite (*i.e.* have both stamens and pistil). Among the commoner plants of the order may be mentioned the spindle-tree, or burn-
ing-bush, as it is sometimes called (*Euonymus*) (Fig. 109, *A*), and the
climbing bitter-sweet (*Celastrus*) (Fig. 109, *D*), belonging to the family
Celastraceæ; the holly and black alder, species of *Ilex*, are examples of
the family *Aquifoliaceæ*; the various species of grape (*Vitis*), the Virginia creeper (*Ampelopsis quinquefolia*), and one or two other cultivated
species of the latter, represent the vine family (*Vitaceæ* or *Ampelidæ*),
and the buckthorn (*Rhamnus*) is the type of the *Rhamnaceæ*.

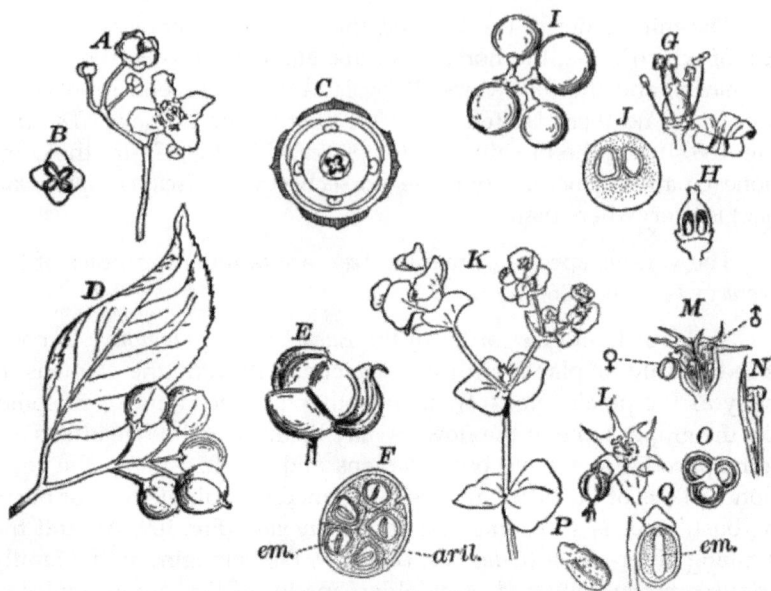

FIG. 109.—*Eucylæ (Frangulinæ), Tricoccæ. A, flowers of spindle-tree, Euony-mus, (Celastraceæ), × 1. B, cross-section of the ovary, × 2. C, diagram of the flower. D, leaf and fruit of bitter-sweet (Celastrus), × ½. E, fruit opening and disclosing the seeds. F, section of a nearly ripe fruit, showing the seeds sur-rounded by the scarlet integument (aril). em. the embryo, × 1. G, flower of grape-vine, Vitis (Vitaceæ), × 2. The corolla has fallen off. H, vertical section of the pistil, × 2. I, nearly ripe fruits of the frost-grape, × 1. J, cross-section of young fruit, × 2. K, a spurge, Euphorbia (Euphorbiaceæ), × ½. L, single group of flowers, surrounded by the corolla-like involucre, × 3. M, section of the same, ♂, male flowers; ♀, female flowers. N, a single male flower, × 5. O, cross-section of ovary, × 6. P, a seed, × 2. Q, longitudinal section of the seed, × 3. em. embryo.*

The fifth group of the *Choripetalæ* is a small one, comprising but a single order (*Tricoccæ*). The flowers are small and inconspicuous, though sometimes, as in some *Euphorbias* and the showy *Poinsettia* of the greenhouses, the leaves or bracts surrounding the inflorescence are conspicuously colored, giving the whole the appearance of a large, showy, single flower. In northern countries the plants are mostly small weeds, of which the various spurges or *Euphorbias* are the most familiar. These plants (Fig. 109, *K*) have the small flowers surrounded by a cup-shaped involucre (*L, M*) so that the whole inflorescence looks like a single flower. In the spurges, as in the other members of the order, the flowers are very simple, being often reduced to a single stamen or pistil (Fig. 109, *M, N*). The plants generally abound in a milky juice which is often poisonous. This juice in a number of tropical genera is the source of India-rubber. Some genera like the castorbean (*Ricinus*) and *Croton* are cultivated for their large, showy leaves.

The water starworts (*Callitriche*), not uncommon in stagnant water, represent the family *Callitrichaceæ*, and the box (*Buxus*) is the type of the *Buxaceæ*.

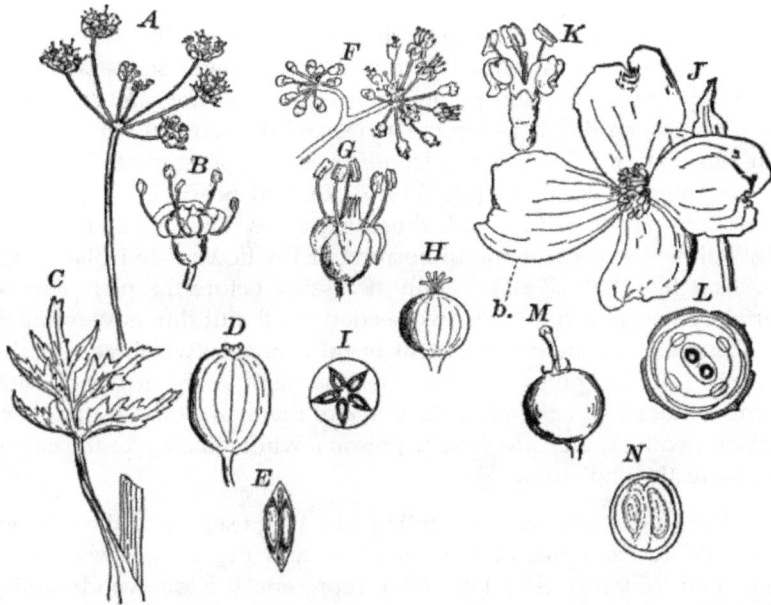

FIG. 110. — *Types of Calycifloræ (Umbellifloræ). A, inflorescence of wild parsnip, Pastinaca (Umbelliferæ), × ½. B, single flower of the same, × 3. C, a leaf, showing the sheathing base, × ¼. D, a fruit, × 2. E, cross-section of D. F, part of the inflorescence of spikenard, Aralia (Araliaceæ), × 1. G, a single flower of the same, × 3. H, the fruit, × 2. I, cross-section of the H. J, inflorescence of dogwood, Cornus (Corneæ). The cluster of flowers is surrounded by four white bracts (b), × ⅓. K, a single flower of the same, × 2. L, diagram of the flower. M, young fruit of another species (Cornus stolonifera) (red osier), × 2. N, cross-section of M.*

The last and highest group of the *Choripetalæ*, the *Calycifloræ*, embraces a very large assemblage of familiar plants, divided into eight orders and thirty-two families. With few exceptions, the floral axis grows up around the ovary, carrying the outer floral leaves above it, and the ovary appears at the bottom of a cup around whose edge the other parts of the flower are arranged. Sometimes, as in the fuchsia, the ovary is grown to the base of the cup or tube, and thus looks as if it were outside the flower. Such an ovary is said to be "inferior" in distinction from one that is entirely free from the tube, and thus is evidently within the flower. The latter is the so-called "superior" ovary. The carpels are usually united into a compound pistil, but may be separate, as in the stonecrop (Fig. 111, *E*), or strawberry (Fig. 114, *C*).

The first order of the *Calyciflorae* (*Umbelliflorae*) has the flowers small, and usually arranged in umbels, *i.e.* several stalked flowers growing from a common point. The ovary is inferior, and there is a nectar-secreting disc between the styles and the stamens. Of the three families, the umbel-worts or *Umbelliferae* is the commonest. The flowers are much alike in all (Fig. 110, *A, B*), and nearly all have large, compound leaves with broad, sheathing bases. The stems are generally hollow. So great is the uniformity of the flowers and plant, that the fruit (Fig. 110, *D*) is generally necessary before the plant can be certainly recognized. This is two-seeded in all, but differs very much in shape and in the development of oil channels, which secrete the peculiar oil that gives the characteristic taste to the fruits of such forms as caraway, coriander, etc. Some of them, like the wild parsnip, poison hemlock, etc., are violent poisons, while others like the carrot are perfectly wholesome.

The wild spikenard (*Aralia*) (Fig. 110, *F*), ginseng, and the true ivy (*Hedera*) are examples of the *Araliaceae*, and the various species of dogwood (*Cornus*) (Fig. 110, *J–N*) represent the dogwood family (*Corneae*).

The second order (*Saxifraginae*) contains eight families, including a number of common wild and cultivated plants. The true saxifrages are represented by several wild and cultivated species of *Saxifraga*, the little bishop's cap or mitre-wort (*Mitella*) (Fig. 111, *D*), and others. The wild hydrangea (Fig. 111, *F*) and the showy garden species represent the family *Hydrangeae*. In these some of the flowers are large and showy, but with neither stamens nor pistils (neutral), while the small, inconspicuous flowers of the central part of the inflorescence are perfect. In the garden varieties, all of the flowers are changed, by selection, into the showy, neutral ones. The syringa or mock orange (*Phila-*

delphus) (Fig. 111, *I*), the gooseberry, and currants (*Ribes*) (Fig. 111, *A*), and the stonecrop (*Sedum*) (Fig. 111, *E*) are types of the families *Philadelpheæ*, *Ribesieæ*, and *Crassulaceæ*.

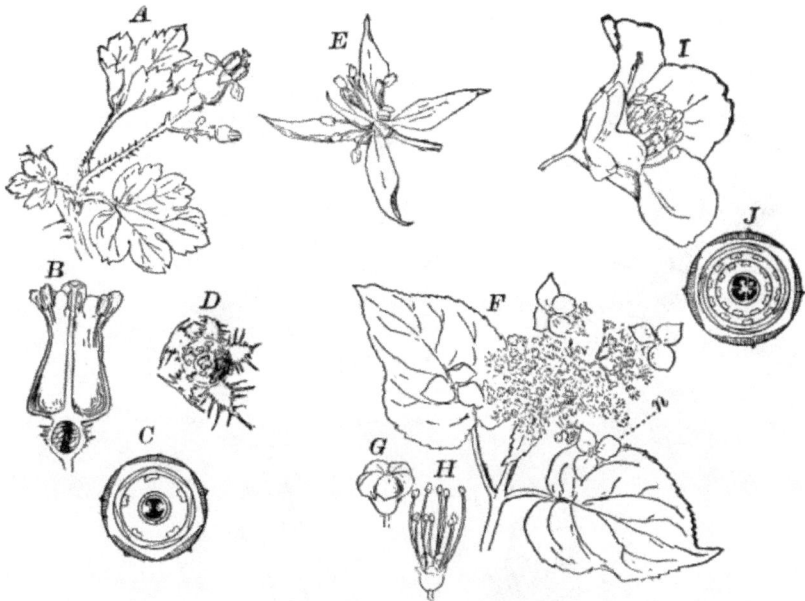

FIG. 111.—*Calycifloræ (Saxifraginæ): A, flowers and leaves of wild gooseberry, Ribes (Ribesieæ), × 1. B, vertical section of the flower, × 2. C, diagram of the flower. D, flower of bishop's-cap, Mitella (Saxifragaceæ), × 3. E, flower of stonecrop, Sedum (Crassulaceæ), × 2. F, flowers and leaves of hydrangea (Hydrangeæ), × ½. n, neutral flower. G, unopened flower, × 2. H, the same, after the petals have fallen away. I, flower of syringa, Philadelphus (Philadelpheæ), × 1. J, diagram of the flower.*

The third order (*Opuntieæ*) has but a single family, the cacti (*Cactaceæ*). These are strictly American in their distribution, and inhabit especially the dry plains of the southwest, where they reach an extraordinary development. They are nearly or quite leafless, and the fleshy, cylindrical, or flattened stems are usually beset with stout spines. The flowers (Fig. 112, *A*) are often very showy, so that many species are cultivated for ornament and are familiar to every one. The beautiful night-blooming cereus, of which there are several species, is one of these. A few species of prickly-pear (*Opuntia*) occur as far north

224

as New York, but most are confined to the hot, dry plains of the south and southwest.

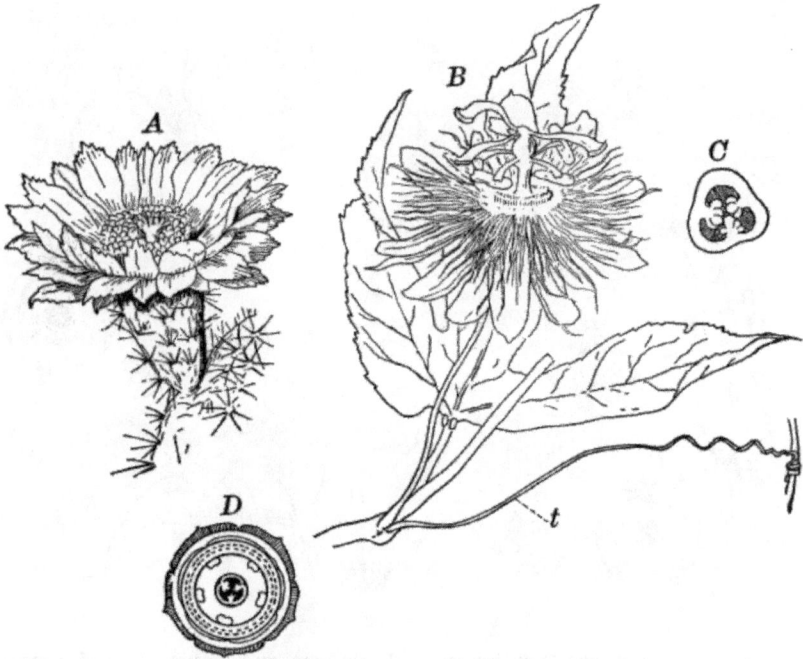

FIG. 112. — *Calyciflorae, Opuntieæ (Passiflorinæ). A, flower of a cactus, Mamillaria (Cactaceæ) (from "Gray's Structural Botany"). B, leaf and flower of a passion-flower, Passiflora (Passifloraceæ), × ½. t, a tendril. C, cross-section of the ovary, × 2. D, diagram of the flower.*

The fourth order (*Passiflorinæ*) are almost without exception tropical plants, only a very few extending into the southern United States. The type of the order is the passion-flower (*Passiflora*) (Fig. 112, *B*), whose numerous species are mostly inhabitants of tropical America, but a few reach into the United States. The only other members of the order likely to be met with by the student are the begonias, of which a great many are commonly cultivated as house plants on account of their fine foliage and flowers. The leaves are always one-sided, and the flowers monœcious. [13] Whether the begonias properly belong with the *Passiflorinæ* has been questioned.

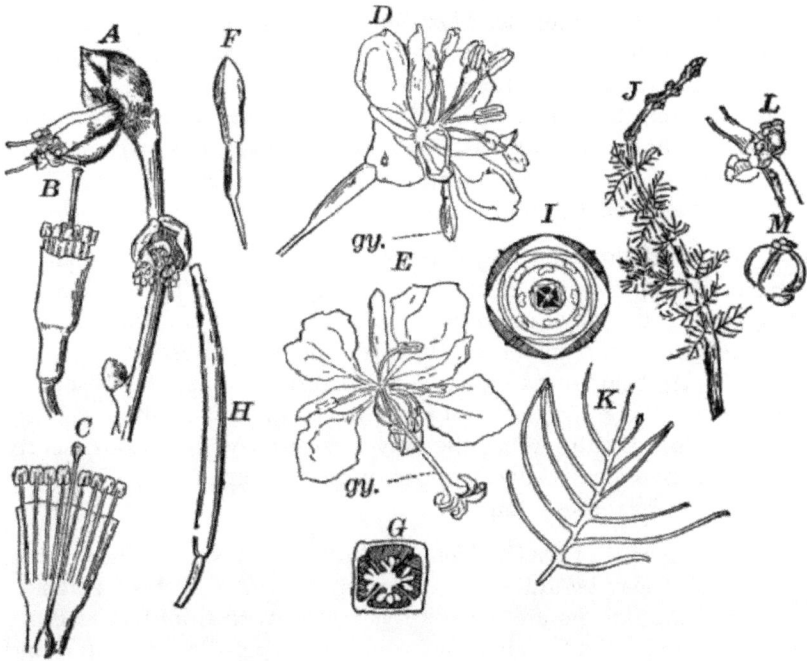

FIG. 113.—*Calycifloræ* (*Myrtifloræ, Thymelinæ*). *A, flowering branch of moosewood, Dirca* (*Thymelæaceæ*), × 1. *B, a single flower,* × 2. *C, the same, laid open. D, a young flower of willow herb, Epilobium* (*Onagraceæ*), × 1. *The pistil* (*gy.*) *is not yet ready for pollination. E, an older flower, with receptive pistil. F, an unopened bud,* × 1. *G , cross-section of the ovary,* × 4. *H, a young fruit,* × 1. *I, diagram of the flower. J, flowering branch of water milfoil, Myriophyllum* (*Haloragidaceæ*), × ½. *K, a single leaf,* × 1. *L, female flowers of the same,* × 2. *M, the fruit,* × 2.

The fifth order (*Myrtifloræ*) have regular four-parted flowers with usually eight stamens, but sometimes, through branching of the stamens, these appear very numerous. The myrtle family, the members of which are all tropical or sub-tropical, gives name to the order. The true myrtle (*Myrtus*) is sometimes cultivated for its pretty glossy green leaves and white flowers, as is also the pomegranate whose brilliant, scarlet flowers are extremely ornamental. Cloves are the dried flower-buds of an East-Indian myrtaceous tree (*Caryophyllus*). In Australia the order includes the giant gum-trees (*Eucalyptus*), the largest of all known trees, exceeding in size even the giant trees of California.

226

Among the commoner *Myrtiflore*, the majority belong to the two families *Onagraceæ* and *Lythraceæ*. The former includes the evening primroses (*Œnothera*), willow-herb (*Epilobium*) (Fig. 113, *D*), and fuchsia; the latter, the purple loosestrife (*Lythrum*) and swamp loosestrife (*Nesæa*). The water-milfoil (*Myriophyllum*) (Fig. 113, *J*) is an example of the family *Haloragidaceæ*, and the *Rhexias* of the eastern United States represent with us the family *Melastomaceæ*.

The sixth order of the *Calyciflore* is a small one (*Thymelinæ*), represented in the United States by very few species. The flowers are four-parted, the calyx resembling a corolla, which is usually absent. The commonest member of the order is the moosewood (*Dirca*) (Fig. 113, *A*), belonging to the first of the three families (*Thymelæaceæ*). Of the second family (*Elæagnaceæ*), the commonest example is *Shepherdia*, a low shrub having the leaves covered with curious, scurfy hairs that give them a silvery appearance. The third family (*Proteaceæ*) has no familiar representatives.

The seventh order (*Rosiflore*) includes many well-known plants, all of which may be united in one family (*Rosaceæ*), with several subfamilies. The flowers are usually five-parted with from five to thirty stamens, and usually numerous, distinct carpels. In the apple and pear (Fig. 114, *I*), however, the carpels are more or less grown together; and in the cherry, peach, etc., there is but a single carpel giving rise to a single-seeded stone-fruit (drupe) (Fig. 114, *E*, *H*). In the strawberry (Fig. 114, *A*), rose (*G*), cinquefoil (*Potentilla*), etc., there are numerous distinct, one-seeded carpels, and in *Spiræa* (Fig. 114, *F*) there are five several-seeded carpels, forming as many dry pods when ripe. The so-called "berry" of the strawberry is really the much enlarged flower axis, or "receptacle," in which the little one-seeded fruits are embedded, the latter being what are ordinarily called the seeds.

FIG. 114.—*Calycifloræ (Rosifloræ). A, inflorescence of strawberry (Fragaria),* × ½. *B, a single flower,* × 1. *C, section of B. D, floral diagram. E, vertical section of a cherry-flower (Prunus),* × 1. *F, vertical section of the flower of Spiræa,* × 2. *G, vertical section of the bud of a wild rose (Rosa),* × 1. *H, vertical section of the young fruit,* × 1. *I, section of the flower of an apple (Pyrus),* × 1. *J, floral diagram of apple.*

From the examples given, it will be seen that the order includes not only some of the most ornamental, cultivated plants, but the majority of our best fruits. In addition to those already given, may be mentioned the raspberry, blackberry, quince, plum, and apricot.

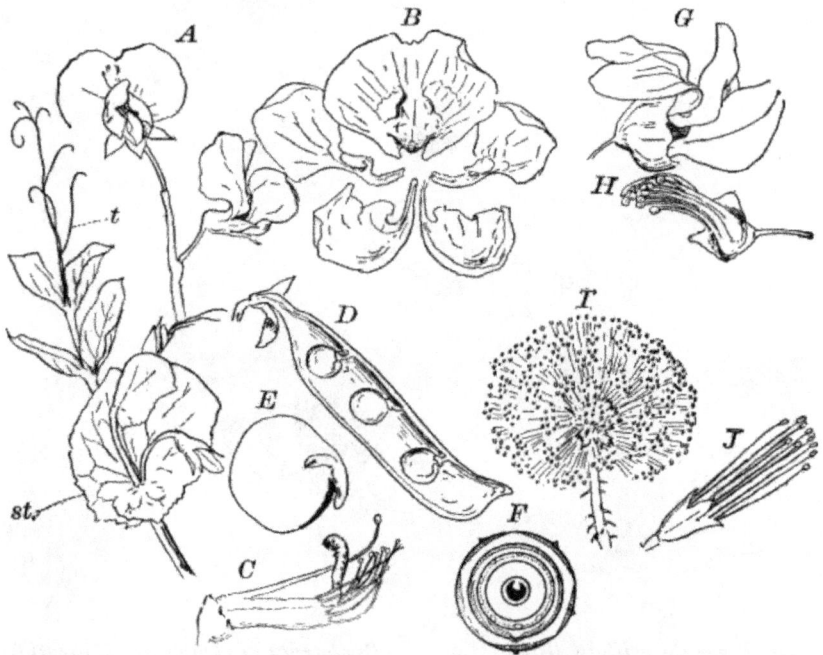

FIG. 115. — *Calycifloræ* (*Leguminosæ*). *A, flowers and leaf of the common pea, Pisum (Papilionaceæ),* × ½. *t, tendril. st. stipules. B, the petals, separated and displayed,* × 1. *C, flower, with the calyx and corolla removed,* × 1. *D, a fruit divided lengthwise,* × ½. *E, the embryo, with one of the cotyledons removed,* × 2. *F, diagram of the flower. G, flower of red-bud, Cercis (Cæsalpinaceæ),* × 2. *H, the same, with calyx and corolla removed. I, inflorescence of the sensitive-brier, Schrankia (Mimosaceæ),* × 1. *J, a single flower,* × 2.

The last order of the *Calycifloræ* and the highest of the *Choripetalæ* is the order *Leguminosæ,* of which the bean, pea, clover, and many other common plants are examples. In most of our common forms the flowers are peculiar in shape, one of the petals being larger than the others, and covering them in the bud. This petal is known as the standard. The two lateral petals are known as the wings, and the two lower and inner are generally grown together forming what is called the "keel" (Fig. 115, *A, B*). The stamens, ten in number, are sometimes all grown together into a tube, but generally the upper one is free from the others (Fig. 115, *C*). There is but one carpel which forms a pod with two valves when ripe (Fig. 115, *D*). The seeds are large, and the embryo fills the seed completely. From the peculiar form of the flower, they are known as *Papilionaceæ* (*papilio,* a butterfly). Many of the *Papilionaceæ* are climbers, either having twining stems, as in the common beans, or else with part of the leaf changed into a tendril as in the pea (Fig. 115, *A*), vetch, etc. The leaves are usually compound.

Of the second family (*Cæsalpineæ*), mainly tropical, the honey locust (*Gleditschia*) and red-bud (*Cercis*) (Fig. 115, *G*) are the commonest examples. The flowers differ mainly from the *Papilionaceæ* in being less perfectly papilionaceous, and the stamens are almost entirely distinct (Fig. 115, *H*). The last family (*Mimosaceæ*) is also mainly tropical. The acacias, sensitive-plant (*Mimosa*), and the sensitive-brier of the southern United States (*Schrankia*) (Fig. 115, *I*) represent this family. The flowers are quite different from the others of the order, being tubular and the petals united, thus resembling the flowers of the *Sympetalæ*. The leaves of *Mimosa* and *Schrankia* are extraordinarily sensitive, folding up if irritated.

CHAPTER XIX.
CLASSIFICATION OF DICOTYLEDONS (Continued).

Division II. — Sympetalæ.

THE *Sympetalæ* or *Gamopetalæ* are at once distinguished from the *Choripetalæ* by having the petals more or less united, so that the corolla is to some extent tubular. In the last order of the *Choripetalæ* we found a few examples (*Mimosaceæ*) where the same thing is true, and these form a transition from the *Choripetalæ* to the *Sympetalæ*.

There are two great divisions, *Isocarpæ* and *Anisocarpæ*. In the first the carpels are of the same number as the petals and sepals; in the second fewer. In both cases the carpels are completely united, forming a single, compound pistil. In the *Isocarpæ* there are usually twice as many stamens as petals, occasionally the same number.

There are three orders of the *Isocarpæ*, viz., *Bicornes*, *Primulinæ*, and *Diospyrinæ*. The first is a large order with six families, including many very beautiful plants, and a few of some economic value. Of the six families, all but one (*Epacrideæ*) are represented in the United States. Of these the *Pyrolaceæ* includes the pretty little pyrolas and prince's-pine (*Chimaphila*) (Fig. 116, *J*); the *Monotropeæ* has as its commonest examples, the curious Indian-pipe (*Monotropa uniflora*), and pine-sap (*M. hypopitys*) (Fig. 116, *L*). These grow on decaying vegetable matter, and are quite devoid of chlorophyll, the former species being pure white throughout (hence a popular name, "ghost flower"); the latter is yellowish. The magnificent rhododendrons and azaleas (Fig. 116, *F*), and the mountain laurel (*Kalmia*) (Fig. 116, *I*), belong to the *Rhodoraceæ*. The heath family (*Ericaceæ*), besides the true heaths (*Erica*, *Calluna*), includes the pretty trailing-arbutus or may-flower (*Epigæa*), *Andromeda*, *Oxydendrum* (Fig. 116, *E*), wintergreen (*Gaultheria*), etc. The last family is represented by the cranberry (*Vaccinium*) and huckleberry (*Gaylussacia*).

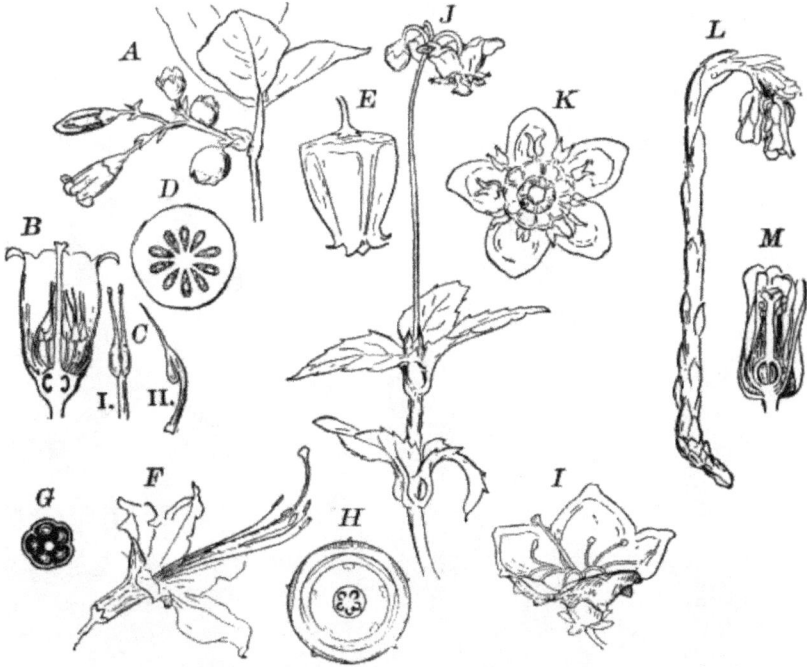

FIG. 116.—*Types of Isocarpous sympetalæ (Bicornes). A, flowers, fruit, and leaves of huckleberry, Gaylussacia (Vacciniex), × 1. B, vertical section of the flower, × 3. C, a stamen: I, from in front; II, from the side, × 4. D, cross-section of the young fruit, × 2. E, flower of sorrel-tree, Oxydendrum (Ericacex), × 2. F, flower of azalea (Rhododendron), × ½. G, cross-section of the ovary, × 3. H, diagram of the flower. I, flower of mountain laurel (Kalmia), × 1. J, prince's-pine, Chimaphila (Pyrolacex), × ½. K, a single flower, × 1. L, plant of pine-sap, Monotropa, (Monotropex), × ½. M, section of a flower, × 1.*

The second order, the primroses (*Primulinæ*), is principally represented in the cooler parts of the world by the true primrose family (*Primulaceæ*), of which several familiar plants may be mentioned. The genus *Primula* includes the European primrose and cowslip, as well as two or three small American species, and the commonly cultivated Chinese primrose. Other genera are *Dodecatheon*, of which the beautiful shooting-star (*D. Meadia*) (Fig. 117, *A*) is the best known. Something like this is *Cyclamen*, sometimes cultivated as a house plant. The moneywort (*Lysimachia nummularia*) (Fig. 117, *D*), as well as other species, also belongs here.

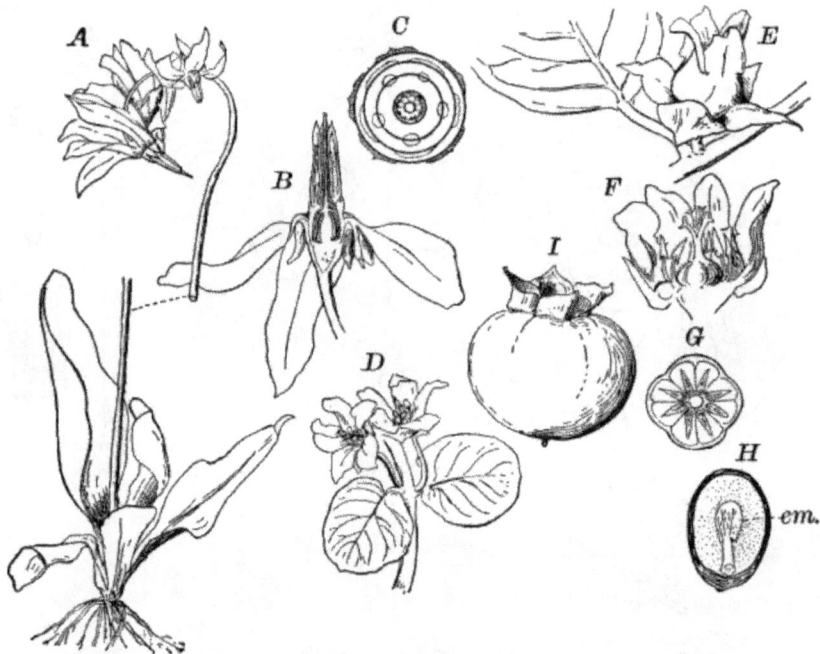

FIG. 117.—*Isocarpous sympetalæ (Primulinæ, Diospyrinæ). A, shooting-star, Dodecatheon (Primulaceæ), × ½. B, section of a flower, × 1. C, diagram of the flower. D, Moneywort, Lysimachia (Primulaceæ), × ½. E, a perfect flower of the persimmon, Diospyros (Ebenaceæ), × 1. F, the same, laid open: section of the young fruit, × 2. H, longitudinal section of a ripe seed, × 1. em. the embryo. I, fruit, × ½.*

The sea-rosemary (*Statice*) and one or two cultivated species of plumbago are the only members of the plumbago family (*Plumbaginæ*) likely to be met with. The remaining families of the *Primulinæ* are not represented by any common plants.

The third and last order of the *Isocarpous sympetalæ* has but a single common representative in the United States; viz., the persimmon (*Diospyros*) (Fig. 117, *E*). This belongs to the family *Ebenaceæ*, to which also belongs the ebony a member of the same genus as the persimmon, and found in Africa and Asia.

The second division of the *Sympetalæ* (the *Anisocarpæ*) has usually but two or three carpels, never as many as the petals. The stamens are also never more than five, and very often one or more are abortive.

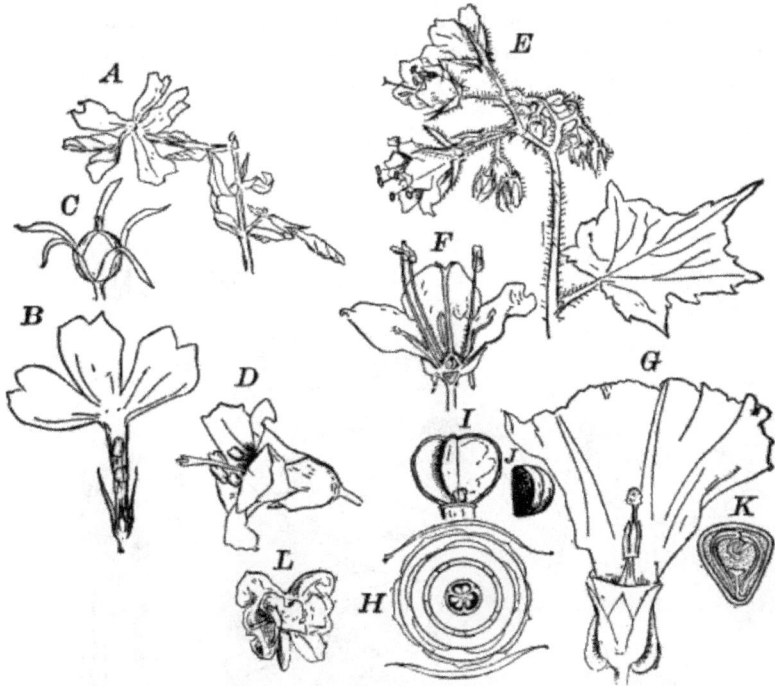

FIG. 118.—*Types of Anisocarpous sympetalæ (Tubifloræ). A, flower and leaves of wild phlox (Polemoniaceæ), × ½. B, section of a flower, × 1. C, fruit, × 1. D, flower of blue valerian (Polemonium), × 1. E, flowers and leaf of water-leaf, Hydrophyllum (Hydrophyllaceæ), × ½. F, section of a flower, × 1. G, flower of wild morning-glory, Convolvulus (Convolvulaceæ), × ½. One of the bracts surrounding the calyx and part of the corolla are cut away. H, diagram of the flower. I, the fruit of a garden morning-glory, from which the outer wall has fallen, leaving only the inner membranous partitions, × 1. J, a seed, × 1. K, cross-section of a nearly ripe seed, showing the crumpled embryo, × 2. L, an embryo removed from a nearly ripe seed, and spread out; one of the cotyledons has been partially removed, × 1.*

The first order (*Tubifloræ*) has, as the name indicates, tubular flowers which show usually perfect, radial symmetry (*Actinomorphism*). There are five families, all represented by familiar plants. The first (*Convolvulaceæ*) has as its type the morning-glory (*Convolvulus*) (Fig. 118, *G*), and the nearly related *Ipomœas* of the gardens. The curious dodder (*Cuscuta*), whose leafless, yellow stems are sometimes very conspicuous, twining over various plants, is a member of this family which has lost its chlorophyll through parasitic habits. The

234

sweet potato (*Batatas*) is also a member of the morning-glory family. The numerous species, wild and cultivated, of phlox (Fig. 118, *A*), and the blue valerian (*Polemonium*) (Fig. 118, *D*), are examples of the family *Polemoniaceæ*.

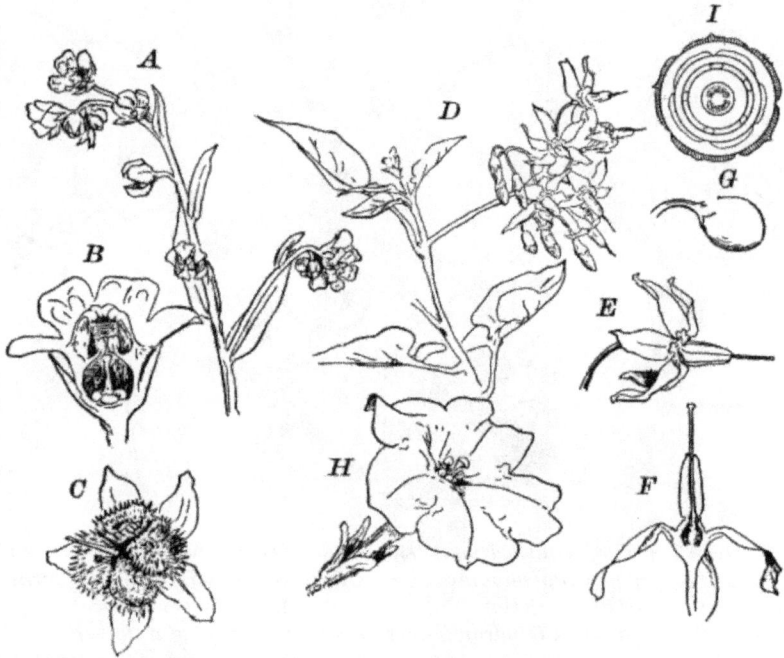

FIG. 119.—*Anisocarpous sympetalæ (Tubifloræ). A, inflorescence of hound's-tongue, Cynoglossum (Borragineæ), × ½. B, section of a flower, × 2. C, nearly ripe fruit, × 1. D, flowering branch of nightshade, Solanum (Solaneæ), × ½. E, a single flower, × 1. F, section of the flower, × 2. G, young fruit, × 1. H, flower of Petunia (Solaneæ), × ½. I, diagram of the flower.*

The third family (*Hydrophyllaceæ*) includes several species of water-leaf (*Hydrophyllum*) (Fig. 118, *E*) and *Phacelia*, among our wild flowers, and species of *Nemophila*, *Whitlavia* and others from the western states, but now common in gardens.

The Borage family (*Borragineæ*) includes the forget-me-not (*Myosotis*) and a few pretty wild flowers, *e.g.* the orange-flowered puccoons (*Lithospermum*); but it also embraces a number of the most troublesome weeds, among which are the hound's-tongue (*Cynoglossum*)

(Fig. 119, *A*), and the "beggar's-ticks" (*Echinospermum*), whose prickly fruits (Fig. 119, *C*) become detached on the slightest provocation, and adhere to whatever they touch with great tenacity. The flowers in this family are arranged in one-sided inflorescences which are coiled up at first and straighten as the flowers expand.

The last family (*Solaneæ*) includes the nightshades (*Solanum*) (Fig. 119, *D*), to which genus the potato (*S. tuberosum*) and the egg-plant (*S. Melongena*) also belong. Many of the family contain a poison-ous principle, *e.g.* the deadly nightshade (*Atropa*), tobacco (*Nicotiana*), stramonium (*Datura*), and others. Of the cultivated plants, besides those already mentioned, the tomato (*Lycopersicum*), and various spe-cies of *Petunia* (Fig. 119, *H*), *Solanum*, and *Datura* are the commonest.

The second order of the *Anisocarpæ* consists of plants whose flowers usually exhibit very marked, bilateral symmetry (*Zygomor-phism*). From the flower often being two-lipped (see Fig. 120), the name of the order (*Labiatifloræ*) is derived.

Of the nine families constituting the order, all but one are repre-sented within our limits, but the great majority belong to two families, the mints (*Labiatæ*) and the figworts (*Scrophularineæ*). The mints are very common and easily recognizable on account of their square stems, opposite leaves, strongly bilabiate flowers, and the ovary split-ting into four seed-like fruits (Fig. 120, *D*, *F*).

The great majority of them, too, have the surface covered with glandular hairs secreting a strong-scented volatile oil, giving the pecu-liar odor to these plants. The dead nettle (*Lamium*) (Fig. 120, *A*) is a thoroughly typical example. The sage, mints, catnip, thyme, lavender, etc., will recall the peculiarities of the family.

The stamens are usually four in number through the abortion of one of them, but sometimes only two perfect stamens are present.

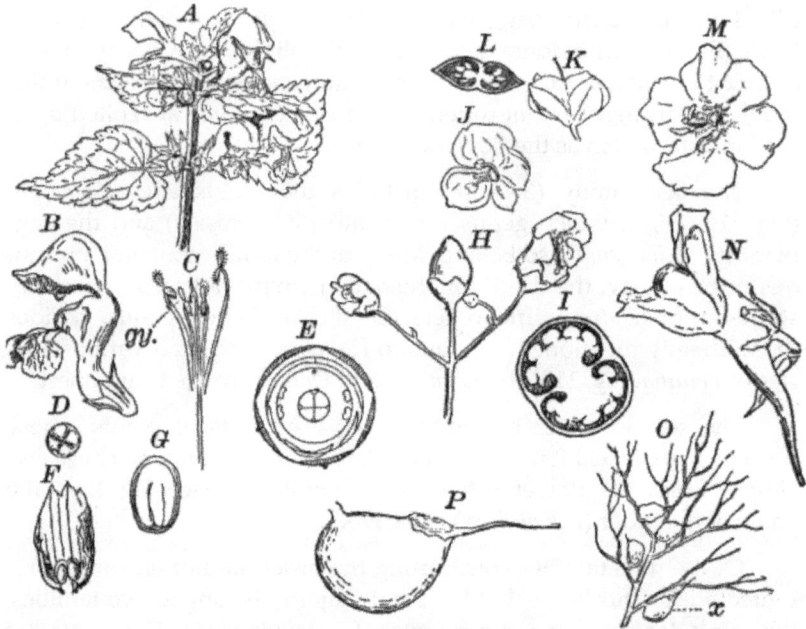

FIG. 120.—*Anisocarpous sympetalæ (Labiatifloræ). A, dead nettle, Lamium, (Labiatæ),* × ½. *B, a single flower,* × 1. *C, the stamens and pistil,* × 1. *D, cross-section of the ovary,* × 2. *E, diagram of the flower; the position of the absent stamen is indicated by the small circle. F, fruit of the common sage, Salvia (Labiatæ),* × 1. *Part of the persistent calyx has been removed to show the four seed-like fruits, or nutlets. G, section of a nutlet,* × 3. *The embryo fills the seed completely. H, part of an inflorescence of figwort, Scrophularia (Scrophularineæ),* × 1. *I, cross-section of the young fruit,* × 2. *J, flower of speedwell, Veronica (Scrophularineæ),* × 2. *K, fruit of Veronica,* × 2. *L, cross-section of K. M, flower of moth-mullein, Verbascum (Scrophularineæ),* × ½. *N, flower of toad-flax, Linaria (Scrophularineæ),* × 1. *O, leaf of bladder-weed, Utricularia (Lentibulariaceæ),* × 1. *x, one of the "traps." P, a single trap,* × 5.

The *Scrophularineæ* differ mainly from the *Labiatæ* in having round stems, and the ovary not splitting into separate one-seeded fruits. The leaves are also sometimes alternate. There are generally four stamens, two long and two short, as in the labiates, but in the mullein (*Verbascum*) (Fig. 120, *M*), where the flower is only slightly zygomorphic, there is a fifth rudimentary stamen, while in others (*e.g. Veronica*) (Fig. 120, *J*) there are but two stamens. Many have large, showy flowers, as in the cultivated foxglove (*Digitalis*), and the native species of *Gerardia*, mullein, *Mimulus*, etc., while a few like the figwort,

Scrophularia (Fig. 120, *H*), and speedwells (*Veronica*) have duller-colored or smaller flowers.

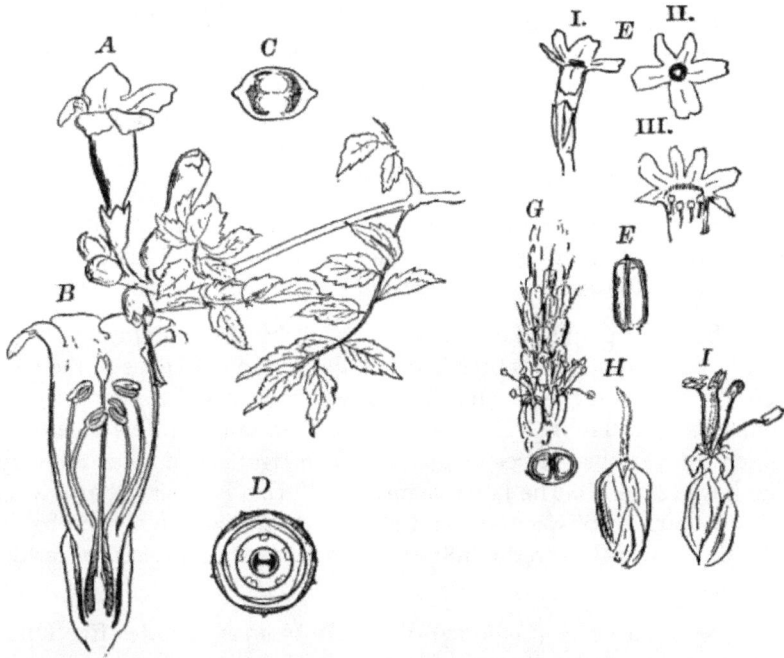

FIG. 121.—*Anisocarpous sympetalæ (Labiatifloræ). A, flowering branch of trumpet-creeper, Tecoma (Bignoniaceæ), × ¼. B, a single flower, divided lengthwise, × ½. C, cross-section of the ovary, × 2. D, diagram of the flower. E, flower of vervain, Verbena (Verbenæ), × 2: I, from the side; II, from in front; III, the corolla laid open. F, nearly ripe fruit of the same, × 2. G, part of a spike of flowers of the common plantain, Plantago (Plantagineæ), × 1; The upper flowers have the pistils mature, but the stamens are not yet ripe. H, a flower from the upper (younger) part of the spike. I, an older expanded flower, with ripe stamens, × 3.*

The curious bladder-weed (*Utricularia*) is the type of the family *Lentibulariaceæ*, aquatic or semi-aquatic plants which possess special contrivances for capturing insects or small water animals. These in the bladder-weed are little sacs (Fig. 120, *P*) which act as traps from which the animals cannot escape after being captured. There does not appear to be here any actual digestion, but simply an absorption of the products of decomposition, as in the pitcher-plant. In the nearly related land form, *Pinguicula*, however, there is much the same arrangement as in the sundew.

The family *Gesneraceæ* is mainly a tropical one, represented in the greenhouses by the magnificent *Gloxinia* and *Achimenes*, but of native plants there are only a few parasitic forms destitute of chlorophyll and with small, inconspicuous flowers. The commonest of these is *Epiphegus*, a much-branched, brownish plant, common in autumn about the roots of beech-trees upon which it is parasitic, and whence it derives its common name, "beech-drops."

The bignonia family (*Bignoniaceæ*) is mainly tropical, but in our southern states is represented by the showy trumpet-creeper (*Tecoma*) (Fig. 121, *A*), the catalpa, and *Martynia*.

The other plants likely to be met with by the student belong either to the *Verbenaceæ*, represented by the showy verbenas of the gardens, and our much less showy wild vervains, also belonging to the genus *Verbena* (Fig. 121, *E*); or to the plantain family (*Plantagineæ*), of which the various species of plantain (*Plantago*) are familiar to every one (Fig. 121, *G*, *I*). The latter seem to be forms in which the flowers have become inconspicuous, and are wind fertilized, while probably all of its showy-flowered relatives are dependent on insects for fertilization.

The third order (*Contortæ*) of the *Anisocarpæ* includes five families, all represented by familiar forms. The first, the olive family (*Oleaceæ*), besides the olive, contains the lilac and jasmine among cultivated plants, and the various species of ash (*Fraxinus*), and the pretty fringe-tree (*Chionanthus*) (Fig. 122, *A*), often cultivated for its abundant white flowers. The other families are the *Gentianaceæ* including the true gentians (*Gentiana*) (Fig. 122, *F*), the buck-bean (*Menyanthes*), the centauries (*Erythræa* and *Sabbatia*), and several other less familiar genera; *Loganiaceæ*, with the pink-root (*Spigelia*) (Fig. 122, *D*), as the best-known example; *Apocynaceæ* including the dog-bane (*Apocynum*) (Fig. 122, *H*), and in the gardens the oleander and periwinkle (*Vinca*).

239

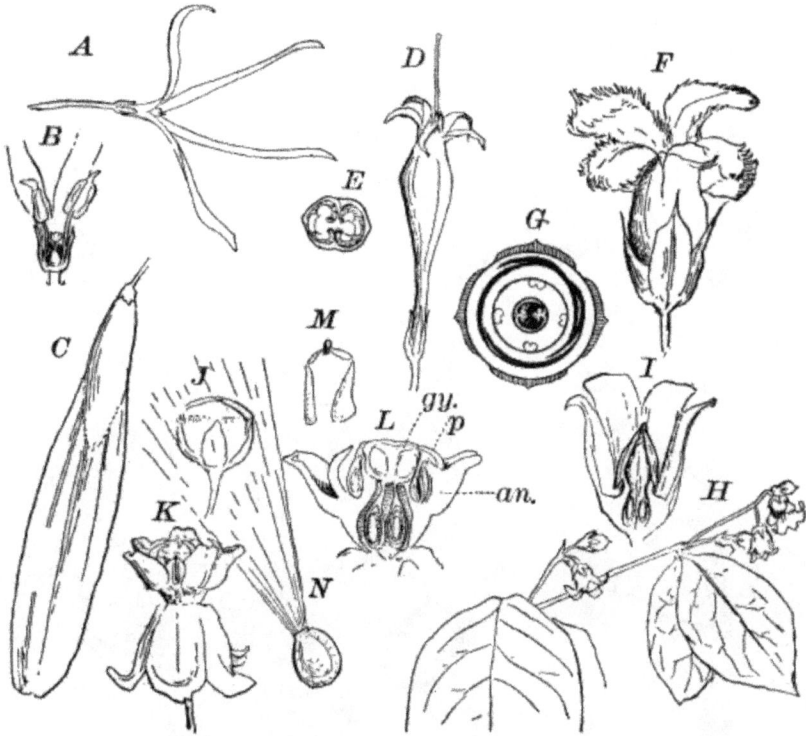

FIG. 122.—*Anisocarpous sympetalæ (Contortæ). A, flower of fringe-tree, Chionanthus (Oleaceæ), × 1. B, base of the flower, with part of the calyx and corolla removed, × 2. C, fruit of white ash, Fraxinus (Oleaceæ), × 1. D, flower of pink-root, Spigelia (Loganiaceæ), × ½. E, cross-section of the ovary, × 3. F, flower of fringed gentian, Gentiana (Gentianaceæ), × ½. G, diagram of the flower. H, flowering branch of dog-bane, Apocynum (Apocynaceæ), × ½. I, vertical section of a flower, × 2. J, bud. K, flower of milk-weed, Asclepias (Asclepiadaceæ), × 1. L, vertical section through the upper part of the flower, × 2. gy. pistil. p, pollen masses. an. stamen. M, a pair of pollen masses, × 6. N, a nearly ripe seed, × 1.*

The last family is the milk-weeds (*Asclepiadaceæ*), which have extremely complicated flowers. Our numerous milk-weeds (Fig. 122, *K*) are familiar representatives, and exhibit perfectly the peculiarities of the family. Like the dog-banes, the plants contain a milky juice which is often poisonous. Besides the true milk-weeds (*Asclepias*), there are several other genera within the United States, but mostly southern in their distribution. Many of them are twining plants and occasionally

240

cultivated for their showy flowers. Of the cultivated forms, the wax-plant (*Hoya*), and *Physianthus* are the commonest.

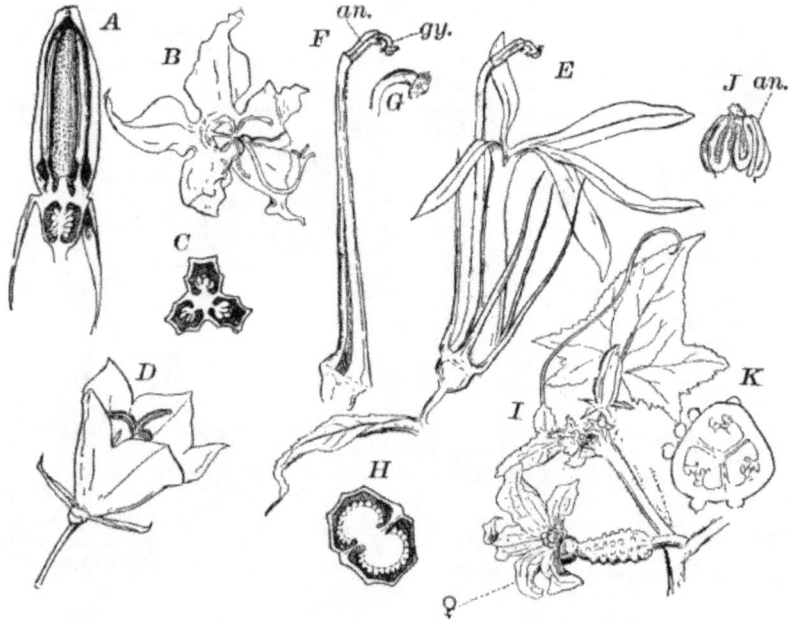

FIG. 123.—*Anisocarpous sympetalæ (Campanulinæ). A, vertical section of the bud of American bell-flower, Campanula (Campanulaceæ), × 2. B, an ex-panded flower, × 1. The stamens have discharged their pollen, and the stigma has opened. C, cross-section of the ovary, × 3. D, flower of the Carpathian bell-flower (Campanula Carpatica), × 1. E, flower of cardinal-flower, Lobelia (Lobeliaceæ), × 1. F, the same, with the corolla and sepals removed. an. the united anthers. gy. the tip of the pistil. G, the tip of the pistil, × 2, showing the circle of hairs surrounding the stigma. H, cross-section of the ovary, × 3. I, tip of a branch of cucumber, Cucurbita (Cucurbitaceæ), with an expanded female flower (♀). J, andrœcium of a male flower, showing the peculiar convo-luted anthers (an.), × 2. K, cross-section of the ovary, × 2.*

The fourth order (*Campanulinæ*) also embraces five families, but of these only three are represented among our wild plants. The bell-flowers (*Campanula*) (Fig. 123, *A, D*) are examples of the family *Cam-panulaceæ*, and numerous species are common, both wild and culti-vated.

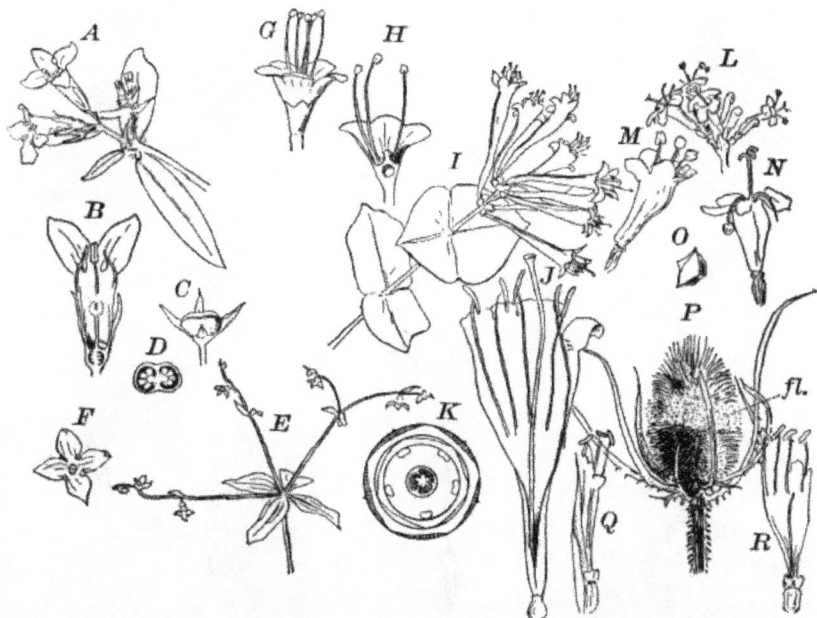

Fig. 124.—*Anisocarpous sympetalæ (Aggregatæ). A, flowering branch of Houstonia purpurea,* × 1 *(Rubiaceæ). B, vertical section of a flower,* × 2. *C, fruit of bluets (Houstonia cœrulea),* × 1. *D, cross-section of the same. E, bedstraw, Galium (Rubiaceæ),* × ½. *F, a single flower,* × 2. *G, flower of arrowwood, Viburnum (Caprifoliaceæ),* × 2. *H, the same, divided vertically. I, flowering branch of trumpet honeysuckle, Lonicera (Caprifoliaceæ),* × ½. *J, a single flower, the upper part laid open,* × 1. *K, diagram of the flower. L, part of the inflorescence of valerian, Valeriana, (Valerianeæ),* × 1. *M, young; N, older flower,* × 2. *O, cross-section of the young fruit; one division of the three contains a perfect seed, the others are crowded to one side by its growth. P, inflorescence of teasel, Dipsacus (Dipsaceæ),* × ¼. *fl. flowers. Q, a single flower,* × 1. *R, the same, with the corolla laid open.*

The various species of *Lobelia*, of which the splendid cardinal-flower (*L. Cardinalis*) (Fig. 123, *E*) is one of the most beautiful, represent the very characteristic family *Lobeliaceæ*. Their milky juice contains more or less marked poisonous properties. The last family of the order is the gourd family (*Cucurbitaceæ*), represented by a few wild species, but best known by the many cultivated varieties of melons, cucumbers, squashes, etc. They are climbing or running plants, and provided with tendrils. The flowers are usually unisexual, sometimes diœcious, but oftener monœcious (Fig. 123, *I*).

242

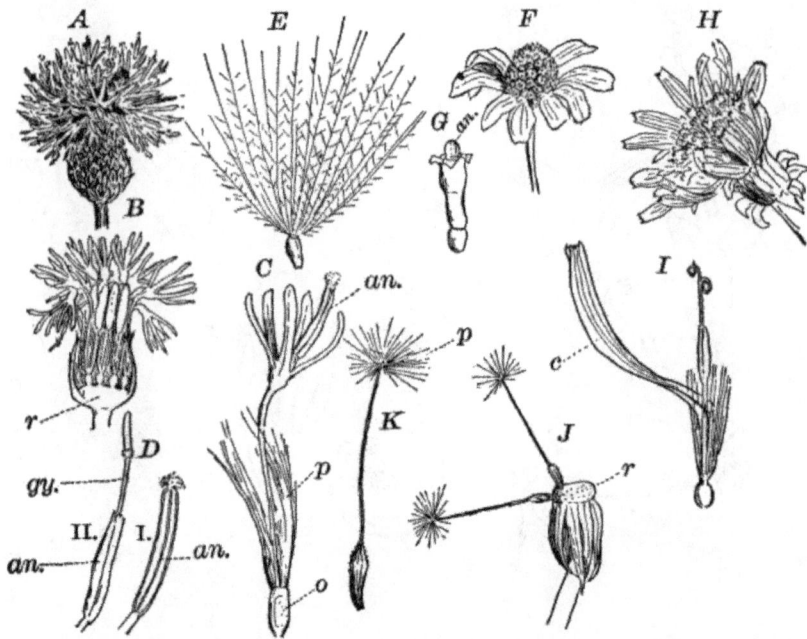

FIG. 125.—*Anisocarpous sympetalæ (Aggregatæ). Types of Compositæ. A, inflorescence of Canada thistle (Cirsium), × 1. B, vertical section of A. r, the receptacle or enlarged end of the stem, to which the separate flowers are attached. C, a single flower, × 2. o, the ovary. p, the "pappus" (calyx lobes). an. the united anthers. D, the upper part of the stamens and pistil, × 3: I, from a young flower; II, from an older one. an. anthers. gy. pistil. E, ripe fruit, × 1. F, inflorescence of may-weed (Maruta). The central part (disc) is occupied by perfect tubular flowers (G), the flowers about the edge (rays) are sterile, with the corolla much enlarged and white, × 2. G, a single flower from the disc, × 3. H, inflorescence of dandelion (Taraxacum), the flowers all alike, with strap-shaped corollas, × 1. I, a single flower, × 2. c, the split, strap-shaped corolla. J, two ripe fruits, still attached to the receptacle (r). The pappus is raised on a long stalk, × 1. K, a single fruit, × 2.*

The last and highest order of the *Sympetalæ*, and hence of the dicotyledons, is known as *Aggregatæ*, from the tendency to have the flowers densely crowded into a head, which not infrequently is closely surrounded by bracts so that the whole inflorescence resembles a single flower. There are six families, five of which have common representatives, but the last family (*Calycereæ*) has no members within our limits.

243

The lower members of the order, *e.g.* various *Rubiaceæ* (Fig. 124, *A*, *E*), have the flowers in loose inflorescences, but as we examine the higher families, the tendency for the flowers to become crowded becomes more and more evident, and in the highest of our native forms *Dipsaceæ* (Fig. 124, *P*) and *Compositæ* (Fig. 125) this is very marked indeed. In the latter family, which is by far the largest of all the angiosperms, including about ten thousand species, the differentiation is carried still further. Among our native *Compositæ* there are three well-marked types. The first of these may be represented by the thistles (Fig. 125, *A*). The so-called flower of the thistle is in reality a close head of small, tubular flowers (Fig. 125, *C*), each perfect in all respects, having an inferior one-celled ovary, five stamens with the anthers united, and a five-parted corolla. The sepals (here called the "pappus") (*p*) have the form of fine hairs. These little flowers are attached to the enlarged upper end of the flower stalk (receptacle, *r*), and are surrounded by closely overlapping bracts or scale leaves which look like a calyx; the flowers, on superficial examination, appear as single petals. In other forms like the daisy and may-weed (Fig. 125, *F*), only the central flowers are perfect, and the edge of the inflorescence is composed of flowers whose corollas are split and flattened out, but the stamens and sometimes the pistils are wanting in these so-called "ray-flowers." In the third group, of which the dandelion (Fig. 125, *H*), chicory, lettuce, etc., are examples, all of the flowers have strap-shaped, split corollas, and contain both stamens and pistils.

The families of the *Aggregatæ* are the following: I. *Rubiaceæ* of which *Houstonia* (Fig. 124, *A*), *Galium* (*E*), *Cephalanthus* (button-bush), and *Mitchella* (partridge-berry) are examples; II. *Caprifoliaceæ*, containing the honeysuckles (*Lonicera*) (Fig. 124, *I*), *Viburnum* (*G*), snowberry (*Symphoricarpus*), and elder (*Sambucus*); III. *Valerianeæ*, represented by the common valerian (*Valeriana*) (Fig. 124, *L*); IV. *Dipsaceæ*, of which the teasel (*Dipsacus*) (Fig. 124, *P*), is the type, and also species of scabious (*Scabiosa*); V. *Compositæ* to which the innumerable, so-called compound flowers, asters, golden-rods, daisies, sunflowers, etc. belong; VI. *Calycereæ*.

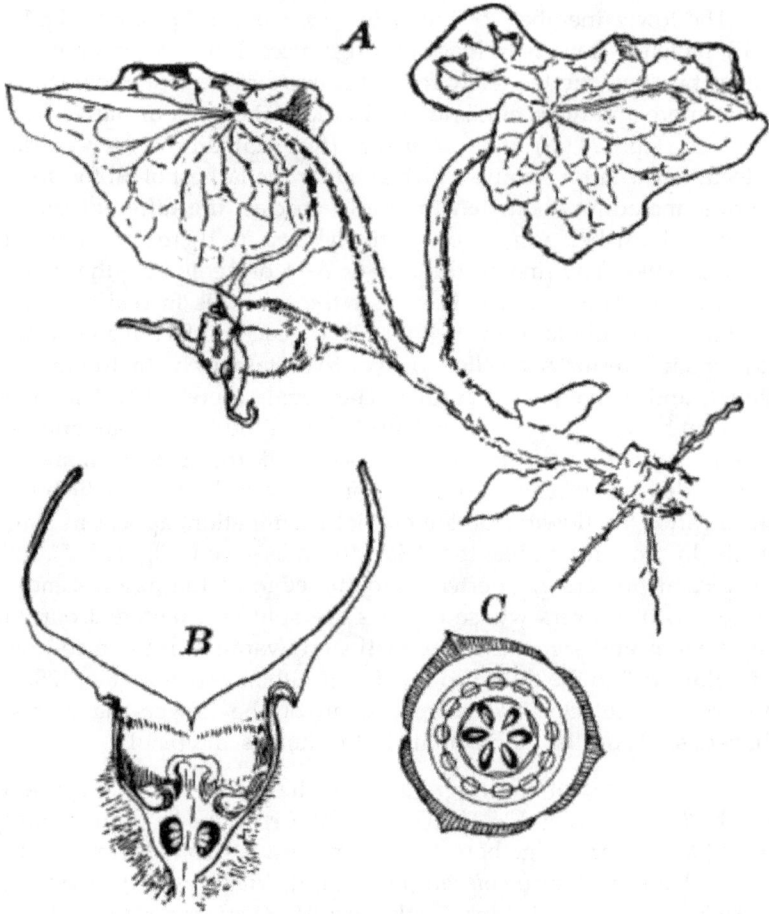

FIG. 126.—*Aristolochiaceæ. A, plant of wild ginger (Asarum)*, × ⅓. *B, vertical section of the flower*, × 1. *C, diagram of the flower.*

Besides the groups already mentioned, there are several families of dicotyledons whose affinities are very doubtful. They are largely parasitic, *e.g.* mistletoe; or water plants, as the horned pond-weed (*Ceratophyllum*). One family, the *Aristolochiaceæ*, represented by the curious "Dutchman's pipe" (*Aristolochia sipho*), a woody twiner with very large leaves, and the common wild ginger (*Asarum*) (Fig. 126), do not appear to be in any wise parasitic, but the structure of their curious flowers differs widely from any other group of plants.

CHAPTER XX.
FERTILIZATION OF FLOWERS.

IF we compare the flowers of different plants, we shall find almost infinite variety in structure, and this variation at first appears to follow no fixed laws; but as we study the matter more thoroughly, we find that these variations have a deep significance, and almost without exception have to do with the fertilization of the flower.

In the simpler flowers, such as those of a grass, sedge, or rush among the monocotyledons, or an oak, hazel, or plantain, among dicotyledons, the flowers are extremely inconspicuous and often reduced to the simplest form. In such plants, the pollen is conveyed from the male flowers to the female by the wind, and to this end the former are usually placed above the latter so that these are dusted with the pollen whenever the plant is shaken by the wind. In these plants, the male flowers often outnumber the female enormously, and the pollen is produced in great quantities, and the stigmas are long and often feathery, so as to catch the pollen readily. This is very beautifully shown in many grasses.

If, however, we examine the higher groups of flowering plants, we see that the outer leaves of the flower become more conspicuous, and that this is often correlated with the development of a sweet fluid (nectar) in certain parts of the flower, while the wind-fertilized flowers are destitute of this as well as of odor.

If we watch any bright-colored or sweet-scented flower for any length of time, we shall hardly fail to observe the visits of insects to it, in search of pollen or honey, and attracted to the flower by its bright color or sweet perfume. In its visits from flower to flower, the insect is almost certain to transfer part of the pollen carried off from one flower to the stigma of another of the same kind, thus effecting pollination.

That the fertilization of a flower by pollen from another is beneficial has been shown by many careful experiments which show that nearly always—at least in flowers where there are special contrivances for cross-fertilization—the number of seeds is greater and the quality better where cross-fertilization has taken place, than where the flower is fertilized by its own pollen. From these experiments, as well as from very numerous studies on the structure of the flower with reference to insect aid in fertilization, we are justified in the conclusion that all bright-colored flowers are, to a great extent, dependent upon insect aid for transferring the pollen from one flower to another, and that many, especially those with tubular or zygomorphic (bilateral) flowers are perfectly incapable of self-fertilization. In a few cases snails have been known to be the conveyers of pollen, and the humming-birds are known in some cases, as for instance the trumpet-creeper (Fig. 121, *A*), to take the place of insects. [14]

At first sight it would appear that most flowers are especially adapted for self-fertilization; but in fact, although stamens and pistils are in the same flower, there are usually effective preventives for avoiding self-fertilization. In a few cases investigated, it has been found that the pollen from the flower will not germinate upon its own stigma, and in others it seems to act injuriously. One of the commonest means of avoiding self-fertilization is the maturing of stamens and pistils at different times. Usually the stamens ripen first, discharging the pollen and withering before the stigma is ready to receive it, *e.g.* willow-herb (Fig. 113, *D*), campanula (Fig. 123, *A, D*), and pea; in the two latter, the pollen is often shed before the flower opens. Not so frequently the stigmas mature first, as in the plantain (Fig. 121, *G*).

In many flowers, the stamens, as they ripen, move so as to place themselves directly before the entrance to the nectary, where they are necessarily struck by any insect searching for honey; after the pollen is shed, they move aside or bend downward, and their place is taken by the pistil, so that an insect which has come from a younger flower will strike the part of the body previously dusted with pollen against the stigma, and deposit the pollen upon it. This arrangement is very beautifully seen in the nasturtium and larkspur (Fig. 99, *J*).

The tubular flowers of the *Sympetalæ* are especially adapted for pollination by insects with long tongues, like the bees and butterflies, and in most of these flowers the relative position of the stamens and pistil is such as to ensure cross-fertilization, which in the majority of them appears to be absolutely dependent upon insect aid.

The great orchid family is well known on account of the singular form and brilliant colors of the flowers which have no equals in these respects in the whole vegetable kingdom. As might be expected, there are numerous contrivances for cross-fertilization among them, some of which are so extraordinary as to be scarcely credible. With few exceptions the pollen is so placed as to render its removal by insects necessary. One of the simpler contrivances is readily studied in the little spring-orchis (Fig. 89) or one of the *Habenarias* (Fig. 90, *G*). In the first, the two pollen masses taper below where each is attached to a viscid disc which is covered by a delicate membrane. These discs are so placed that when an insect enters the flower and thrusts its tongue into the spur of the flower, its head is brought against the membrane covering the discs, rupturing it so as to expose the disc which adheres firmly to the head or tongue of the insect, the substance composing the disc hardening like cement on exposure to the air. As the insect withdraws its tongue, one or both of the pollen masses are dragged out and carried away. The action of the insect may be imitated by thrusting a small grass-stalk or some similar body into the spur of the flower, when on withdrawing it, the two pollen masses will be removed from the flower. If we now examine these carefully, we shall see that they change position, being nearly upright at first, but quickly bending downward and forward (Fig. 89, *D*, II, III), so that on thrusting the stem into another flower the pollen masses strike against the sticky stigmatic surfaces, and a part of the pollen is left adhering to them.

The last arrangement that will be mentioned here is one discovered by Darwin in a number of very widely separated plants, and to which he gave the name "heterostylism." Examples of this are the primroses (*Primula*), loosestrife (*Lythrum*), partridge-berry (*Mitchella*), pickerel-weed (*Pontederia*), (Fig. 84, *I*), and others. In these there are two, sometimes three, sets of flowers differing very much in the relative lengths of stamens and pistil, those with long pistils having short stamens and *vice versa*. When an insect visits a flower with short stamens, that part is covered with pollen which in the short-styled (but long-stamened) flower will strike the stigma, as the pistil in one flower is almost exactly of the length of the stamens in the other form. In such flowers as have three forms, *e.g. Pontederia*, each flower has two different lengths of stamens, both differing from the style of the same flower. Microscopic examination has shown that there is great variation in the size of the pollen spores in these plants, the large pollen from the long stamens being adapted to the long style of the proper flower.

It will be found that the character of the color of the flower is related to the insects visiting it. Brilliantly colored flowers are usually visited by butterflies, bees, and similar day-flying insects. Flowers opening at night are usually white or pale yellow, colors best seen at night, and in addition usually are very strongly scented so as to attract the night-flying moths which usually fertilize them. Sometimes dull-colored flowers, which frequently have a very offensive odor, are visited by flies and other carrion-loving insects, which serve to convey pollen to them.

Occasionally, flowers in themselves inconspicuous are surrounded by showy leaves or bracts which take the place of the petals of the showier flowers in attracting insect visitors. The large dogwood (Fig. 110, *J*), the calla, and Jack-in-the-pulpit (Fig. 86, *A*) are illustrations of this.

CHAPTER XXI.
HISTOLOGICAL METHODS.

IN the more exact investigations of the tissues, it is often necessary to have recourse to other reagents than those we have used hitherto, in order to bring out plainly the more obscure points of structure. This is especially the case in studies in cell division in the higher plants, where the changes in the dividing nucleus are very complicated.

For studying these the most favorable examples for ready demonstration are found in the final division of the pollen spores, especially of some monocotyledons. An extremely good subject is offered by the common wild onion (*Allium Canadense*), which flowers about the last of May. The buds, which are generally partially replaced by small bulbs, are enclosed in a spathe or sheath which entirely conceals them. Buds two to three millimetres in length should be selected, and these opened so as to expose the anthers. The latter should now be removed to a slide, and carefully crushed in a drop of dilute acetic acid (one-half acid to one-half distilled water). This at once fixes the nuclei, and by examining with a low power, we can determine at once whether or not we have the right stages. The spore mother cells are recognizable by their thick transparent walls, and if the desired dividing stages are present, a drop of staining fluid should be added and allowed to act for about a minute, the preparation being covered with a cover glass. After the stain is sufficiently deep, it should be carefully withdrawn with blotting paper, and pure water run under the cover glass.

The best stain for acetic acid preparations is, perhaps, gentian violet. This is an aniline dye readily soluble in water. For our purpose, however, it is best to make a concentrated, alcoholic solution from the dry powder, and dilute this as it is wanted. A drop of the alcoholic solution is diluted with several times its volume of weak acetic acid (about two parts of distilled water to one of the acid), and a drop of this mixture added to the preparation. In this way the nucleus alone is stained and is rendered very distinct, appearing of a beautiful violet-blue color.

If the preparation is to be kept permanently, the acid must all be washed out, and dilute glycerine run under the cover glass. The preparation should then be sealed with Canada balsam or some other cement, but previously all trace of glycerine must be removed from the slide and upper surface of the cover glass. It is generally best to gently wipe the edge of the cover glass with a small brush moistened with alcohol before applying the cement.

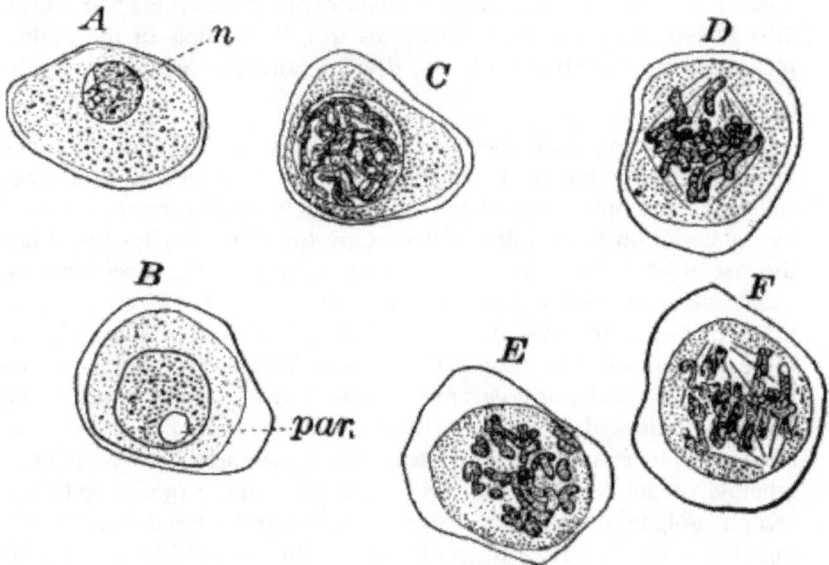

FIG. 127.—*A, pollen mother cell of the wild onion. n, nucleus. B–F, early stages in the division of the nucleus. par. nucleolus; acetic acid, gentian violet, × 350.*

If the spore mother cells are still quite young, we shall find the nucleus (Fig. 127, *A*, *n*) comparatively small, and presenting a granular appearance when strongly magnified. These granules, which appear isolated, are really parts of filaments or segments, which are closely twisted together, but scarcely visible in the resting nucleus. On one side of the nucleus may usually be seen a large nucleolus (called here, from its lateral position, paranucleus), and the whole nucleus is sharply separated from the surrounding protoplasm by a thin but evident membrane.

The first indication of the approaching division of the nucleus is an evident increase in size (*B*), and at the same time the colored granules become larger, and show more clearly that they are in lines indicating the form of the segments. These granules next become more or less confluent, and the segments become very evident, appearing as deeply stained, much-twisted threads filling the nuclear cavity (Fig. 127, C), and about this time the nucleolus disappears.

The next step is the disappearance of the nuclear membrane so that the segments lie apparently free in the protoplasm of the cell. They arrange themselves in a flat plate in the middle of the cell, this plate appearing, when seen from the side, as a band running across the middle of the cell. (Fig. 127, D, shows this plate as seen from the side, E seen from above.)

About the time the nuclear plate is complete, delicate lines may be detected in the protoplasm converging at two points on opposite sides of the cell, and forming a spindle-shaped figure with the nuclear plate occupying its equator. This stage (D), is known as the "nuclear spindle." The segments of the nuclear plate next divide lengthwise into two similar daughter segments (F), and these then separate, one going to each of the new nuclei. This stage is not always to be met with, as it seems to be rapidly passed over, but patient search will generally reveal some nuclei in this condition.

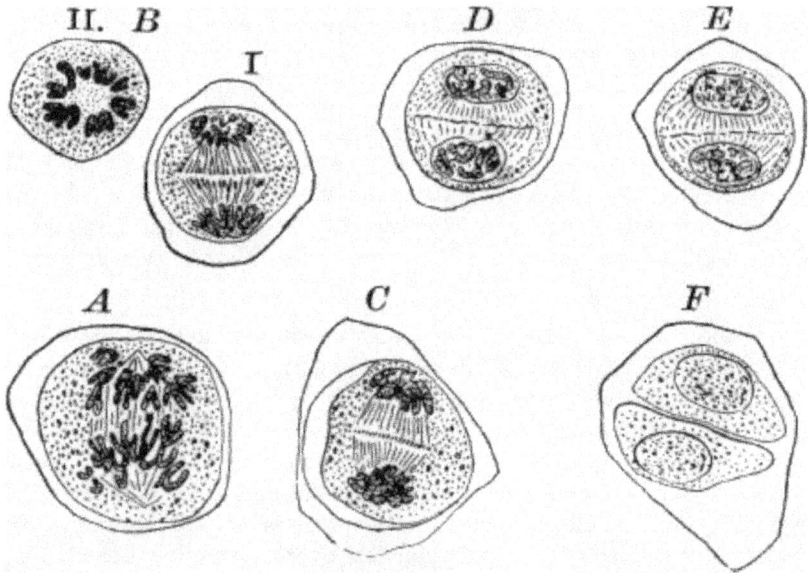

Fig. 128.—Later stages of nuclear divisions in the pollen mother cell of wild onion, × 350. All the figures are seen from the side, except B ii, which is viewed from the pole.

Although this is almost impossible to demonstrate, there are probably as many filaments in the nuclear spindle as there are segments (in this case about sixteen), and along these the nuclear segments travel slowly toward the two poles of the spindle (Fig. 128, A, B). As the two sets of segments separate, they are seen to be connected by very numerous, delicate threads, and about the time the young nuclei reach the poles of the nuclear spindle, the first trace of the division wall appears in the form of isolated particles (microsomes), which arise first as thickenings of these threads in the middle of the cell, and appear in profile as a line of small granules not at first extending across the cell, but later, reaching completely across it (Fig. 128, C, E). These granules constitute the young cell wall or "cell plate," and finally coalesce to form a continuous membrane (Fig. 128, F).

The two daughter nuclei pass through the same changes, but in reverse order that we saw in the mother nucleus previous to the formation of the nuclear plate, and by the time the partition wall is complete the nuclei have practically the same structure as the first stages we examined (Fig. 128, *F*). [15]

This complicated process of nuclear division is known technically as "karyokinesis," and is found throughout the higher animals as well as plants.

The simple method of fixing and staining, just described, while giving excellent results in many cases, is not always applicable, nor as a rule are the permanent preparations so made satisfactory. For permanent preparations, strong alcohol (for very delicate tissues, absolute alcohol, when procurable, is best) is the most convenient fixing agent, and generally very satisfactory. Specimens may be put directly into the alcohol, and allowed to stay two or three days, or indefinitely if not wanted immediately. When alcohol does not give good results, specimens fixed with chromic or picric acid may generally be used, and there are other fixing agents which will not be described here, as they will hardly be used by any except the professional botanist. Chromic acid is best used in a watery solution (five per cent chromic acid, ninety-five per cent distilled water). For most purposes a one per cent solution is best; in this the objects remain from three or four to twenty-four hours, depending on size, but are not injured by remaining longer. Picric acid is used as a saturated solution in distilled water, and the specimen may remain for about the same length of time as in the chromic acid. After the specimen is properly fixed it must be thoroughly washed in several waters, allowing it to remain in the last for twenty-four hours or more until all trace of the acid has been removed, otherwise there is usually difficulty in staining.

As staining agents many colors are used. The most useful are hæmatoxylin, carmine, and various aniline colors, among which may be mentioned, besides gentian violet, safranine, Bismarck brown, methyl violet. Hæmatoxylin and carmine are prepared in various ways, but are best purchased ready for use, all dealers in microscopic supplies having them in stock. The aniline colors may be used either dissolved in alcohol or water, and with all, the best stain, especially of the nucleus, is obtained by using a very dilute, watery solution, and allowing the sections to remain for twenty-four hours or so in the staining mixture.

Hæmatoxylin and carmine preparations may be mounted either in glycerine or balsam. (Canada balsam dissolved in chloroform is the ordinary mounting medium.) In using glycerine it is sometimes necessary to add the glycerine gradually, allowing the water to slowly evaporate, as otherwise the specimens will sometimes collapse owing to the too rapid extraction of the water from the cells. Aniline colors, as a rule, will not keep in glycerine, the color spreading and finally fading entirely, so that with most of them the specimens must be mounted in balsam.

Glycerine mounts must be closed, which may be done with Canada balsam as already described. The balsam is best kept in a wide-mouthed bottle, specially made for the purpose, which has a glass cap covering the neck, and contains a glass rod for applying the balsam.

Before mounting in balsam, the specimen must be completely freed from water by means of absolute alcohol. (Sometimes care must be taken to bring it gradually into the alcohol to avoid collapsing. [16]) If an aniline stain has been used, it will not do to let it stay more than a minute or so in the alcohol, as the latter quickly extracts the stain. After dehydrating, the specimen should be placed on a clean slide in a drop of clove oil (bergamot or origanum oil is equally good), which renders it perfectly transparent, when a drop of balsam should be dropped upon it, and a perfectly clean cover glass placed over the preparation. The chloroform in which the balsam is dissolved will soon evaporate, leaving the object embedded in a transparent film of balsam between the slide and cover glass. No further treatment is necessary. For the finer details of nuclear division or similar studies, balsam mounts are usually preferable.

It is sometimes found necessary in sectioning very small and delicate organs to embed them in some firm substance which will permit sectioning, but these processes are too difficult and complicated to be described here.

The following books of reference may be recommended. This list is, of course, not exhaustive, but includes those works which will probably be of most value to the general student.

- 1. GOEBEL. Outlines of Morphology and Classification.

- 2. SACHS. Physiology of Plants.

- 3. DE BARY. Comparative Anatomy of Ferns and Phanerogams.

- 4. DE BARY. Morphology and Biology of Fungi, Mycetozoa, and Bacteria.

These four works are translations from the German, and take the place of Sachs's Text-book of Botany, a very admirable work published first about twenty years ago, and now somewhat antiquated. Together they constitute a fairly exhaustive treatise on general botany. — New York, McMillan & Co.

- 5. GRAY. Structural Botany. — New York, Ivison & Co.

- 6. GOODALE. Physiological Botany. — New York, Ivison & Co.

These two books cover somewhat the same ground as 1 and 2, but are much less exhaustive.

- STRASBURGER. Das Botanische Practicum. — Jena.

Where the student reads German, the original is to be preferred, as it is much more complete than the translations, which are made from an abridgment of the original work. This book and the next (7 and 8) are laboratory manuals, and are largely devoted to methods of work.

- 7. ARTHUR, BARNES, and COULTER. Plant Dissection. — Holt & Co., New York.

- 8. WHITMAN. Methods in Microscopic Anatomy and Embryology. — Casino & Co., Boston.

For identifying plants the following books may be mentioned: —

- Green algæ (exclusive of desmids, but including *Cyanophyceæ* and *Volvocineæ*).

- WOLLE. Fresh-water Algæ of the United States. — Bethlehem, Penn.

- Desmids. WOLLE. Desmids of the United States. — Bethlehem, Penn.

- The red and brown algæ are partially described in FARLOW'S New England Algæ. Report of United States Fish Commission, 1879. — Washington.

- The *Characeæ* are being described by Dr. F. F. ALLEN of New York. The first part has appeared.

- The literature of the fungi is much scattered. FARLOW and TRELEASE have prepared a careful index of the American literature on the subject.

- Mosses. LESQUEREUX and JAMES. Mosses of North America. — Boston, Casino & Co.

- BARNES. Key to the Genera of Mosses. — Bull. Purdue School of Science, 1886.

- Pteridophytes. UNDERWOOD. Our Native Ferns and their Allies. — Holt & Co., New York.

- Spermaphytes. GRAY. Manual of the Botany of the Northern United States. 6th edition, 1890. This also includes the ferns, and the liverworts. — New York, Ivison & Co.

- COULTER. Botany of the Rocky Mountains. — New York, Ivison & Co.

- CHAPMAN. Flora of the Southern United States. — New York, 1883.

- WATSON. Botany of California.

FOOTNOTES.

[1] For the mounting of permanent preparations, see Chapter XIX.

[2] The term "colony" is, perhaps, inappropriate, as the whole mass of cells arises from a single one, and may properly be looked upon as an individual plant.

[3] Algæ (sing. *alga*).

[4] "Host," the plant or animal upon which a parasite lives.

[5] The antheridia, when present, arise as branches just below the oögonium, and become closely applied to it, sometimes sending tubes through its wall, but there has been no satisfactory demonstration of an actual transfer of the contents of the antheridium to the egg cell.

[6] The filaments are attached to the surface of the leaf by suckers, which are not so readily seen in this species as in some others. A mildew growing abundantly in autumn on the garden chrysanthemum, however, shows them very satisfactorily if a bit of the epidermis of a leaf on which the fungus is just beginning to grow is sliced off with a sharp razor and mounted in dilute glycerine, or water, removing the air with alcohol. These suckers are then seen to be globular bodies, penetrating the outer wall of the cell (Fig. 40).

[7] Sing. *soredium*.

[8] Sing. *basidium*.

[9] A vessel differs from a tracheid in being composed of several cells placed end to end, the partitions being wholly or partially absorbed, so as to throw the cells into close communication.

[10] In most conifers the symmetrical form of the young tree is maintained as long as the tree lives.

[11] See the last chapter for details.

[12] The three outer stamens are shorter than the inner set.

[13] Monœcious: having stamens and carpels in different flowers, but on the same plant.

[14] In a number of plants with showy flowers, *e.g.* violets, jewelweed, small, inconspicuous flowers are also formed, which are self-fertilizing. These inconspicuous flowers are called "cleistogamous."

[15] The division is repeated in the same way in each cell so that ultimately four pollen spores are formed from each of the original mother cells.

[16] For gradual dehydrating, the specimens may be placed successively in 30 per cent, 50 per cent, 70 per cent, 90 per cent, and absolute alcohol.